# Operational Amplifiers

# Operational Amplifiers

ROBERT G. SEIPPEL

RESTON PUBLISHING COMPANY, INC.
*A Prentice-Hall Company*
Reston, Virginia

**Library of Congress Cataloging in Publication Data**

Seippel, Robert G.
Operational amplifiers.



Bibliography.
Includes index.
1. Operational amplifiers. I. Title.
TK7871.58.06S44 1983 621.381'735 82-12253
ISBN 0-8359-5242-8


*A Prentice-Hall Company*
Reston, Virginia 22090



10 9 8 7 6 5 4 3 2 1

PRINTED IN THE UNITED STATES OF AMERICA

# Contents

# Preface

The operational amplifier integrated circuit (IC) has become a workhorse in modern solid-state circuitry. Its uses and variations in design are almost unlimited. Such information immediately instills in many the fear that anything this versatile must also be complex. The purpose of this book is to remove the mystery that may surround operational amplifier application.

The author wishes to acknowledge the fine operational amplifier books available on the market. The best of these are listed in the bibliography. Some of them have been referenced in the text.

What, then, makes this book different from others, and why is it unique?

The book begins with an introduction to electricity—electronics and integrated-circuit operational amplifiers. Enough history is supplied in the first dozen or so pages to stimulate the reader to study more about the marvelous background of the electrical/electronic field.

Chapter 2 provides an integrated-circuit hierarchy; it sets the stage for operational amplifier studies and lets the reader know just where op amps stand in this massive IC chain of command.

Chapter 3 reviews the entire fabrication process. This is probably the first op amp book (or any other) that discusses the fabrication process from raw silicon to the finished solid-state device. A pictorial display of the Czochralski method of silicon crystal growing has been provided by Siltec Corporation. Planar fabrication process data were supplied by Analog Devices.

Operational amplifier fundamentals are examined in Chapter 4. These are much the same as you would find in any op amp book.

Chapter 5 presents preliminary design information. Special considerations—such as how to choose an op amp and how to deal with minus signs in design—are covered.

Chapter 6 contains the meat of the book. It begins with a description of the operational amplifier circuit, then describes its peculiar characteristics and its operation. Description is followed by design procedures and examples. Most basic circuits are covered in detail. Each application has a circuit; each circuit is succeeded by a step-by-step design procedure. Immediately following the procedure is a design example in "cookbook" style. Test verification is provided as a final topic. The entire process is fundamental but thorough.

Chapters 7 and 8 are slanted toward instrumentation. Since instrumentation engineers and technicians utilize the bridge and the op amp, the author believes these chapters are required. Further, instrumentation and isolation amplifiers are outgrowths of the op amp, directed toward that need in instrumentation.

Finally, Chapter 9 describes a large selection of op amp circuits, from amplifiers to tachometers. This will provide the reader with a variety of instant applications.

The author believes this book can fill several voids. First, it can furnish the experienced designer with instant applications, thereby saving precious time. Second, it can provide all interested electronic-minded people with a do-it-yourself tool for operational amplifier design. Finally, it can provide instructors at all levels of learning with student-centered instructional material that does not require preparation.

It is recommended that this book be used in post solid-state device courses as an interface between courses on the transistor and those on integrated circuits. However, this suggestion is not necessary for understanding. With the rapidly expanding world of solid-state electronics, people do not have time to wade through a lot of rhetoric to arrive at basic objectives.

*ACKNOWLEDGMENTS*: No writer can develop a technical book without the support of many people in industry. When readers see a new book containing new and innovative ideas or techniques, you can rest assured that the technical data were derived from industry. Research leads us all to the originators of processes and applications.

For instance, Chapter 3 of this book describes the process of developing a solid-state device from raw silicon to finished product. The author traveled to Siltec Corporation in Menlo Park, California to see the process for himself. This chapter would not be possible had it not been for the support of Siltec Corporation and Susan Phinney. Likewise, a trip to Analog Devices in Massachusetts and the support of J. Eric Janson led to the writing of another part of Chapter 3, Chapter 7, and Chapter 8. National Semiconductor provided data for Chapter 9. I'd like to thank Steve Sields and Scott Foote for their support.

Dave Howell of IC Masters and Clive McCarthy of Fairchild Camera and Instrument Corporation, Linear Division, were helpful in developing the Appendices. I'd also like to thank an old-timer by the name of Hugo Gernsback. Hugo is the father of today's electronic publications and electronic apparatus. In 1908 he published Volume I of *Modern Electrics*, which led to *Radio Craft Magazine*, *Radio Electronics*, and *Popular Electronics*. He also supplied amateurs of the day with apparatus for their experiments. Without Hugo Gernsback, we would not be nearly as informed as we are about this electronic world of ours.

In a final tribute, I'd like to thank Charles Barnes and Jo Anne Bly for the fine artwork, my editor Patricia Rayner at Reston, and my loving wife Hazel for her typing and general all-around everything.

ROBERT G. SEIPPEL

# References & Permissions

CHAPTER 3:

Fabrication process information in this chapter was provided in part by Eric Janson, Analog Devices Corporation, Norwood, Massachusetts. Wafer manufacturing process information in this chapter was provided by Anthony Bonora, Vice President of Research, Siltec Corporation, Menlo Park, California.

CHAPTER 7:

The author highly recommends the Transducer Interfacing Handbook, a book written by a number of engineers with Analog Devices Corporation and edited by Daniel H. Sheingold. It is a handbook to be used as a guide to Analog Signal Conditioning. Bridge networks are, of course, an integral part of signal conditioning.

Illustrations in this chapter are reprinted with permission of Analog Devices, Norwood, Massachusetts. All copyrights © are reserved.

CHAPTER 8:

Illustrations in this chapter are reprinted with permission of Analog Devices, Norwood, Massachusetts. All copyrights © are reserved.

# Operational Amplifiers

# 1
# Introduction to Electronics

The history of electronics probably started when people noticed the spark-producing properties of amber, as this material was rubbed on cloth or fur. Around 600 B.C. Thales, the Greek, discovered that mysterious sparks resulted when he rubbed the mineral amber. The Greek word for amber is *electrum*, which serves as the root for the word "electricity."

The lodestone also interested the ancients, who believed that its magnetic potency was destroyed by garlic. This myth lasted for many centuries; it was mentioned as late as 1544 in a treatise on physics written by Philip Melanchton, a German Protestant reformer. Similar theories arose concerning amber and diamonds. The most notable of these was that iron rubbed by a diamond became a magnet, and that diamonds, when rubbed, would attract bits of paper and particles of dust.

Dr. William Gilbert (1540–1603), an English scientist, called such ideas "the chattering of barbers"; instead of accepting such talk at face value, he set out to disprove the myths and in so doing, became the "Father of Electricity."

Gilbert compiled a huge list of materials that could be charged. These included such items as "true jewels, sulphur, sealing wax, rock salt, alum, and resin." He assigned the title "electric" to the phenomenon, and a new era in science came into being. Practically everything he rubbed turned "electric," and this ultimately led Gilbert to invent the electroscope, which he used to test his materials. The electroscope consisted of a straw that was pivoted like a compass needle to indicate the approach of a charged body. Gilbert also conceived of the earth as a huge magnet with magnetic poles and a field of force about it. This laid a positive foundation for many future scientific discoveries.

Sir Thomas Browne (1604–1682), English physicist and author, performed many experiments with the lodestone and magnetism. The first of these refuted the possibility of magnetism losing its potency when subjected to garlic juice. By far the most important of Browne's experiments was his "wireless communication." Using an idea previously conceived by John Baptista Porta, Browne employed two compasses with letters written around their perimeters. He imagined that if the two needles were magnetized together, then separated, the turning of one would cause the other to follow to the same position. Letters selected could make words. If this happened, it would mean that remote communication could take place. However, the second compass needle remained at North—according to Browne, it stayed "like the Pillars of Hercules." Nevertheless, this first attempt at remote communications inspired many more people of the day to continue to work on the idea.

Otto von Guericke (1602–1686), a German burgomaster, was famous for the "Magdeburg" experiment, in which he proved atmospheric pressure. It was his experiments with static electricity that were important to the electronics field, however. In 1672 he constructed an electrical generating device made of a globe of sulphur mounted on an axle and turned by a crank. The globe was rubbed by the hand when it was rotated and after some rubbing became electrified (statically charged). Although charging had taken place, von Guericke noticed, particles would be attracted and remain for a while, then be repelled. The particles were assuming a like charge to that of the sulphur, then they were repelled because, as we know today, "like charges repel." Here he was laying the foundation for what we know to be fact. In a further experiment von Guericke repelled a feather from his sulphur globe and chased it around the room with the globe in his arms. The feather was repelled by the electron stream from a lit candle, which reversed its charge. When the feather flew back to the globe, von Guericke was observing electronic emission without knowing it. Three centuries passed before anyone knew more of this phenomenon. Von Guericke had also observed electricity in a static condition, but left the study of dynamic electricity to Ben Franklin and his kite.

Stephen Gray (1695–1736), an Englishman, studied the effects of charged bodies being transmitted. He discovered the effects of materials we know as conductors and insulators.

About the same time, in France, Paris Dufay (1698–1739) experimented with Gray's discoveries. He found that metal wire or wet objects were the best conductors, though the most difficult to electrify. He also found that the best insulators were easy to electrify. Dufay built a quarter-mile-long line, held up by glass tubes and wetted down. Then he transmitted a charged impluse from one end to the other. This finding was probably the first transmission and consequently an important step in electricity. Historians also suggest that this was the first time wire was used as a conductor.

In these times, scientists and experimenters were still playing around with friction apparatus, since no means for generating electricity had been found. In 1745, E. G. von Kleist of Germany invented the first condenser. Later the same year, Peter van Musschenbroek, of Leyden, Holland, built a Leyden jar (a more publicized condenser) with the help of his assistant Cunaeus. Musschenbroek believed that electricity could be bottled up and saved for later use. His idea was to charge water up in a jar, place a cork on it, and, since the glass was a good insulator, save the charge for the future. A circuit was made of a gun barrel, a friction machine, and a brass wire which entered a jar half-filled with water. Musschenbroek cranked the friction machine while his assistant held the jar in one hand and tried to draw off sparks with the other. If Cunaeus had placed the jar on the table, nothing would have happened. The improvised condenser finally charged up, and Cunaeus was in the middle of it all. The spark that resulted was tremendous, and Cunaeus' body provided a good path for electron flow. We can assume that Cunaeus, shocked by the whole situation, took the next few days off with pay.

Another scientist of the day, Nollet, a French Abbé (1700–1770), heard about the Leyden jar experiment and duplicated it in a big way. He had 200 soldiers stand hand in hand in a circle and placed a severe charge through them. To his delight, all 200 soldiers jumped at the same time. This pleased Nollet immensely; the soldiers were not so happy.

Both Musschenbroek and Nollet wanted to find out what caused this effect—but were lacking in volunteers to repeat the experiment too many times. They found that when the jar was placed on a table, this nullified the experiment. In further experiments, the jar was placed on a metal-plate base on the table. Later, an outside tinfoil cover was substituted, and thus the condenser was born.

Benjamin Franklin (1706–1790) made history in electricity with his famous kite and lightning experiment in 1751. Franklin was a statesman, philosopher, and, by no means least, a scientist. It was he who discovered the principle of conservation of electrical charge, establishing that there are positive- and negative-type charges. He further related lightning and sparks obtained from electrically charged bodies that become discharged. He invented the lightning rod to prevent damage caused by lightning. Probably the best-known theory of this man is that "like charges repel, unlike charges attract."

Luigi Galvani (1737–1798) was an Italian professor of anatomy. Up to his time, there were only two known methods for creating electricity: by means of a frictional machine, and by means of lightning and the clouds, as in Franklin's experiment. In 1780 Galvani noticed that an electrical charge applied to a dead frog's leg made the leg twitch. He further noted that several dead frogs he had suspended on a metal balcony responded to a lightning storm with some motion before the storm. Later, he determined that any two metals joined together would cause the twitching if attached to the frog in two separate places. Galvani reasoned that the electricity was a fluid which made a circuit from muscles to the nerves of the frog then through the metal conductor and back to the muscle. He called this fluid "animal electricity." Galvani thought the electricity originated in the frog. We know it now as "galvanic" electricity caused by two dissimilar metals in contact.

Alessandro Volta (1745–1827), an Italian professor, invented the *voltaic pile*. It was simply a pile of alternately stacked zinc and copper discs separated by a water-wet pasteboard. Volta discovered that a distinct shock was achieved when the fingertips were placed on each end of the pile. He also noted that when the watered pasteboard dried out, the shock diminished. He then placed his zinc and copper discs in separate jars containing a light acid. This was the first battery (storage device) known to man. It was of course a great help to future scientists. Volta's name became immortal when the electrical unit of force, the *volt*, was named after him.

A follow-up modification was made on the principle of the voltaic pile by English experimenters William Nicholson (1753–1815) and Sir Anthony Carlisle (1769–1840). They found that water could be decomposed into its elements of hydrogen and oxygen when an electrical current was passed through it. Hence we had the discovery of *water electrolysis*, as we know it today.

Hans Christian Oersted (1777–1851) was a Danish physics professor who for thirteen years had experimented with electricity and its effects on the compass needle. Oersted had read Franklin's reports on the effects and relations of the two. While lecturing to a class in 1820, he noticed that a compass needle wavered when a switch was closed on a circuit to the voltaic pile. He dismissed his class, closed the switch, and positioned the compass at different points on the wire. Along the wire the compass needle deflected; across the wire, the needle came to a stationary position. Above the wire, the needle went to the opposite direction than when it was below the wire. Oersted had determined the basis for establishing magnetic lines of force, thus providing the fundamentals for measuring and indicating electrical instruments.

In the same year, reportedly only a week after Oersted made his discovery, André Marie Ampere (1775–1836), a French physicist, found that two parallel wires carrying electricity and free to move, attract each other if the current flows in the same direction but repel each other if the current flows in opposing directions. Ampere also found that the wire carrying an electrical current would attract a magnetized needle and that the needle attracted the wire as well. The basic unit of current, the *ampere*, was named in this scientist's honor.

In the same year, 1820, French physicist François Arago (1786–1854) developed the first electromagnet by placing a current through a coil of wire.

In 1826, the basic law for all circuit analysis was born. George Simon Ohm (1789–1854), a Bavarian physicist, had analyzed current flow in relation to voltage and resistance. The law that bears his name is this: "A current flowing in any closed circuit is proportional to the force or voltage and inversely proportional to the resistance of the wire." Today we express this law simply as

$$I = \frac{E}{R}, \quad \text{where } I = \text{current}, \quad E = \text{voltage}, \quad \text{and} \quad R = \text{resistance}$$

Joseph Henry was an American physicist (1797–1878) who improved on the electromagnet developed by Arago in 1820. He used silk-covered wire which allowed the use of many layers of turns. This development of insulated wire (1836) permitted him to make electromagnets large enough to lift several pounds. The electrical unit of inductance is called the *henry* to honor this scientist.

The name of Samuel F. B. Morse (1791–1872), an American artist, is important in the history of electrical discoveries. Morse created the telegraph system (1832–1837) and conceived a code (the Morse code) which permitted the transmission and reception of messages. The Morse code is still used in telegraphy.

Michael Faraday (1791–1867), an Englishman, was one of the most productive of all scientists. His first studies in electricity dealt with the basic principle of the electric motor. Faraday reasoned that if a current in a wire caused a magnetized needle to rotate, then a magnet should cause a wire carrying current to do likewise. Faraday formulated the laws of induction, which allowed him later to build the first electric generator. In fact, it was he who discovered alternating current. He also created the first transformer and developed the dielectric constant from his study of condensers. Faraday was probably one of the most important and productive scientists that ever lived.

The "electromagnetic theory of light" was mathematically elaborated upon by the Scottish physicist, James Clerk Maxwell (1831–1879), though the thought was conceived by Faraday. According to this theory, light, electric waves, and magnetic waves of varying frequency travel in the same medium (namely, ether). Since ether permeates all matter, a current may exist in and about a conductor but is essentially guided by it.

In 1865, Dr. Mahlon Loomis, an American dentist, conducted some of the first experiments in transmitting and receiving messages. He reasoned that the earth's atmosphere could be used as a conductor, since he had read somewhere that the atmosphere is continuously charged with electricity. To test his theory, Loomis sent up two kites from high mountains in West Virginia, 18 miles apart. The kites were covered with large squares of copper screen, and the connecting cord had fine copper wire within. The wire from each kite string was connected to one side of a galvanometer. The other side of the galvanometer was held in readiness by Loomis so he could establish a connection to a coil buried in the earth. The receiving station was connected between meter and earth coil. The transmitting end held by Loomis was grounded, and the galvanometer at the receiving station actually dipped.

In 1875, the microphone (magnetic transducer) was invented by Alexander Graham Bell (1847–1922). Bell, of course, was a major contributor to electronics. Two years later, D. E. Hughes, an Englishman, invented the carbon microphone.

In 1885, Sir William H. Preece and A. W. Heaviside, both Englishmen, sent signals to each other over a distance of 1,000 yards. They placed two telegraph lines parallel, with a telephone receiver on the receiving end. The telegraph signals could be clearly heard in the phone receiver without actual connection between the two lines—this was due to what is known now as induction. In lay language, the phenomenon is called "cross-talk."

Heinrich Hertz (1857–1894), a German physicist, shares the title "Father of Radio" with Marconi. Indeed, for his work in the field we use his name to designate cycles per second. Radio waves are also called Hertzian waves. Hertz was extremely interested in Maxwell's theories on light, magnetism, and electrical waves. In an attempt to learn more, Hertz made the the first spark transmitter and receiver. The transmitter consisted of a Leyden jar and a coil of wire, the ends of which were left open so that a small gap was formed. For the receiver, he employed a similar coil, with a gap arrangement, located in the opposite end of the room. When the Leyden jar was sufficiently charged, it discharged through the gap in the wire coil and the oscillating waves generated were launched in the air of the room and swept across the receiving coil, causing sparks to jump the gap in the receiving coil. Hertz measured the velocity of these waves and found that they were the same velocity as light (186,000 miles per second). He also measured their length and therefore verified Maxwell's theories.

Nikola Tesla (1856–1943) was a Serbo-American inventor. In 1893 he invented what we know as the Tesla coil. His idea was to use the earth as a conductor and create stationary electrical waves on it. His coil (really a broadband transmitter) created high-frequency oscillations of a broad nature. Tesla made no effort to detect the oscillations, and allowed that golden opportunity to go to Marconi. He did, however, develop the receiving capability later, some five years after Marconi.

Guglielmo Marconi (1874–1937) was the Italian physicist who designed the wireless telegraph. Considered by some scientists of the time as an interloper, Marconi nevertheless had the designer's insight to put together knowledge and devices of the time to create a workable wireless telegraph. He combined the coherer (invented by Brauley in 1892), the Russian Popoff's decoherer, the Hertzian theories, and a number of scientific writings to create his telegraph. In fact, when his first patent was granted in England in 1896 for wireless telegraphy, the apparatus differed only slightly from that developed by his predecessors. Marconi was a master designer, and after his first British patent he made rapid strides. He was only 23 years old when he formed the Wireless Telegraph and Signal Company. His first major accomplishment was transmission and reception between warships 12 miles apart. In 1899 he adapted Sir Oliver Lodge's principle of syntony (tuning of circuits), which eliminated much of the interference in transmission, and obtained a patent for it in 1900. Also in 1899, he designed longer-distance units that were used in ship-to-shore installations. When Marconi was asked by the *New York Herald* to cover the international yacht race, he accepted, and the American people thus knew who the winners were before the yachts arrived ashore. Marconi wanted to sell the telegraph to the U. S. Navy, but it rejected his contract terms. His famous transatlantic transmission of the letter S was probably the high point of his design career.

Two Americans, Reginald A. Fessendon and Lee De Forest, are co-bearers of the title "Father of American Radio." Fessendon was very much aware of Marconi's work. He himself had experimented with the wireless for years with the same devices. Fessendon knew that Marconi's design was adapted only to dampened wave transmission; thus it would not tolerate superimposing on its voice or other irregular waves. Consequently, Fessendon began experiments in continuous-wave (cw) transmission. In these efforts he invented the arc transmitter. Since the coherer would not work with the oscillating wave produced by the transmitter, Fessendon created the electrolytic detector (the forerunner of the diode), which allowed current to flow in only one direction. The detector consisted of a solution of acid and water with a fine silver wire inserted. This device was a hugh improvement over the coherer. Later Fessendon conceived of the idea of employing an alternator that produced sine waves greater than 120 cycles per second. This was truly an advance, because it eliminated the spark gaps and arcs and laid the foundation for carrier waves.

Meanwhile, Lee De Forest was experimenting with wireless, and in 1901 he created a wireless less cumbersome and more efficient than Marconi's. For this he used the electrolytic detector, but Fessendon claimed rights to this; there was a court battle, and De Forest lost.

Around 1904, J. Ambrose Fleming, an Englishman, developed the two-element valve (the diode) while working for Marconi. Fleming had been a consultant to the Edison Electric Light Company of America, and remembered what was called the "Edison effect." He believed that this phenomenon could be of use to detect radio waves, and he was right.

Shortly after the two-element valve came into being, the greatest device known to electronics—the Audion—was born. The Audion was simply a Fleming two-element tube with an inserted grid element. Yet this device was so revolutionary that it must be considered the epitome of all electronic-device inventions. Every amplifier device we have now has indeed followed the same concepts.

Legal problems plagued the Fleming and De Forest inventions. Marconi claimed that the Audion tube was a Fleming valve with a grid interposed; the patent was therefore Fleming's, as Marconi claimed, so De Forest could not manufacture the tubes. De Forest owned the Audion patent, and therefore Marconi could not fabricate the Audion.

Around the same time (1907), the status of the tube as a detector was encroached upon by the crystal detector. G. W. Pickard created his first detector from silicon. Later detectors were made of galena, iron pyrites, and Carborundum. The crystal detector was especially useful in detecting feeble signals and irregular pulses such as modulated continuous waves. The crystal detector was inexpensive and also increased the interest of the public in radio.

It was in 1912 that De Forest and associates Logwood and Van Ethan discovered the amplifying properties of the Audion by placing it in cascade (one stage after another). This work interested AT&T. Since De Forest had had problems with finances, he sold the rights to the Audion tube to AT&T for a sum that was embarrassing even in those days.

But there were problems with the Audion. The early Audion was a soft tube, and any plate voltage above 30 volts dc caused it to turn blue because of gas ionization. Gas modules interfered with the electron flow. In effect, the Audion's problems didn't help much in the competition with the solid crystal detector. However, the problems were overcome by Irving Langmuir, a physicist at GE laboratories who discovered a process for creating high vacuum in about 1912. Almost simultaneously, Western Electric (AT&T) did the same thing.

About the same time, the idea of feedback came into focus. De Forest and his associates had developed regenerative feedback to increase sensitivity in the Audion. However, Edwin Armstrong, a student at Columbia at the time, sought a patent for feedback and De Forest was back in court once more. The courts ruled for De Forest, but the patent didn't become his until 1934.

As a result of the hard vacuum tube and AT&T's work, De Forest triodes were used in 1914 to send messages around the world. World War I, of course, spurred activity in the radio field, and after the war hard tubes were in such demand that everyone started making them, patent or no patent.

It must be remembered that with the advancement of the tube, other devices such as the transformer, inductors, condensers, wire, and the manufacturing processes also underwent changes for the better. It also must be stated that the methods for hooking up the circuitry, such as the Hartley and Colpitts oscillators, the Armstrong superheterodyne receiver, and Lacault's Ultradyne, contributed greatly to electronics.

At this point the author must apologize. This text is directed toward a chronological history leading to linear integrated circuits. It is difficult to cover the entire history of electronics in this short chapter. We shall have to leave many historical events for the reader to research.

From this point onward, devices were advanced so quickly that it is difficult to determine just who created what and in what order. The number of inventions increased with the population and the wealth of the world, especially the United States.

Probably the first major tube available from a major supplier was the Westinghouse VT series (VT1, VT2, etc.). This was used in regenerative circuits and operated from 22.5 vdc for plate voltage. It was the best of the detectors available in 1917.

The RCA company was formed in 1919 and bought out the Marconi Company of America. It began to make the Marconi (Fleming) tube and warned others against infringing on the patent. In 1921, RCA released the UV200 (a detector) and UV201 (an amplifier). Both were triodes with brass shells. Both had heating filaments. At that time AT&T was manufacturing with De Forest's patent. AT&T made a deal in 1920 to exchange privileges with RCA to keep them both out of court. The UV201A (with a thoriated filament) was developed soon after. This new tube required only a small amperage for its filament; it was very expensive and not very durable, which led to the overhaul of tubes and tube reactivators.

The next advance in tube manufacture was in low-current devices, which used only 60 milliamps of current. The B eliminator became popular in 1926, along with the Raytheon helium filamentless (cold tube). These were both direct heater tubes.

In 1927, the indirect heater came into being. This increased the interest in ac as the heater voltage. Battery-operated tubes were being pushed out of the picture except in isolated areas where no ac was available.

In 1928, the screen grid tube was released by several companies. This had tremendous amplifying power, and was stable and sensitive. In fact, it was such a good amplifier that it caused stations to interfere with each other. The defect, called cross-modulation, was a worry to engineers until 1931, when variable mu tubes and the pentode tube arrived on the scene.

In 1933, the first multipurpose tube, the pentode, was released. This was sensitive, and still could be driven by a detector with high-power output. Also in 1933, the 6.3 vac heater voltage was developed, along with the acorn tube, as a result of ultrahigh-frequency studies.

Metal tubes were created in 1935; they made higher amplification possible. The electron multiplier and secondary emission tubes followed soon after. Then came the cathode ray, the iconscope, the beam-power tube, and the cold cathode. Further advancements were phototubes and thryatroms. Tubes are still being used in the 1980s, and it doesn't look as though they will disappear in the near future.

For years, since the crystal detector and even further back, scientists had been working on solutions to eliminate the frailties of the tube. In 1948, at Bell Laboratories, this problem was somewhat alleviated. The transistor was born, thanks to Dr. William Shockley, Dr. John Bardeen, and Dr. Walter Brattain, who received a Nobel prize for their efforts.

It must be assumed that solid-state devices have been around since the advent of the piezoelectrical devices, the crystal quartz, Rochelle salts, barium titanate lead metaniobate, and lead titanate zirconate. This does not mean that the term "solid-state device" has two meanings. Quite often the term is used synonymously with "semiconductor devices." Solid state is any electrical device that has no moving parts or gaps.

The word "semiconductor" implies something that is neither a great conductor nor an insulator. In electronics, the semiconductor is usually an atom that has four electrons in its outer valence. These are primarily silicon and germanium. Quite often other atoms such as arsenic, phosphorous, and boron are also lumped into the general category of semiconductor. These are dopant materials used for the fabrication of semiconductor devices. These atoms have three or five electrons in their valence, one less or one more than the silicon atom.

In any event, William Shockley of Bell Telephone Laboratories conceived the whole idea for the transistor around 1940. It wasn't until 1947 that his colleagues Brattain and Bardeen succeeded in creating the transistor. The device was a low-frequency amplifier that used a piece of silicon immersed in a salt solution. Then, in June 1948, the first transistor was created using germanium and three pointed wires. Two of the wires were connected to external power and the third to a battery providing bias the same as the grid in a vacuum tube.

In 1949, Shockley published the now-famous article "The Theory of pn Junctions in Semiconductors and pn Junction Transistors" in the *Bell System Technical Journal.* The devices used in experimentation were not very good. The idea of plunging wire into the material was problematical, because the crystal was hard and did not allow uniformity. A single crystal material was suggested by Shockley and the first of the commercial devices were made in 1951. Unwanted impurities plagued the manufacturers until 1954, when William G. Pfann, a metallurgist at Bell Labs, found a method of refining the silicon to purity. By 1955, excellent quality transistors existed. Then, later in 1955, the diffusion technique was developed and the transistor was on its way. Leaders in fabrication techniques were Western Electric Company, Texas Instruments, RCA, GE, Westinghouse, and Raytheon.

Between 1955 and 1960 Dr. Leo Esaki, a Japanese physicist, in the course of his work for General Electric, created a gallium arsenide transistor that could handle frequencies of 4,000 to 5,000 megacycles. The types of transistors made since that time until the development of integrated circuits are almost uncountable. The transistor, of course, led the way to miniaturization and later microelectronics.

The transistor dominated the solid-state device world until 1958. It was in that year that the integrated circuit was conceived and constructed by Jack Kilby of Texas Instruments. The circuit was the first monolithic device.

Chapters 2 and 3 provide information on the integrated circuit hierarchy and the fabrication process. We leave the history of the integrated circuit to the reader, for he is actually experiencing that history. We shall, however, take a cursory look at the history of the operational amplifier.

In a paper written by John R. Ragazzini entitled "Analysis of Problems in Dynamics," the words "operational amplifier" were coined. The article was printed by the IRE in May 1947 and is a classic in electronics. The paper defined the work of the National Defence Research Council (1943/44) and described the properties of the op amp. It also covered the work of a technical aide, George A. Philbrick, during these times. Philbrick helped develop the first plug-in operational amplifier using vacuum tubes.

The first modular solid-state operational amplifiers were introduced by Burr-Brown Research Corp. and G. A. Philbrick Researches, Inc. in 1962. Since that time the op amp has been the workhorse of linear systems.

The first op amp that was provided to the public (or accepted by the public) was Fairchild Semiconductor's $\mu$A702. The device had low common mode rejection (CMR), low voltage gain, and used ridiculous supply voltages such as +12 and −6 vdc. Furthermore, it tended to burn out when temporarily shorted. The $\mu$A702 was produced in 1963. Despite these problems, the device was the best in its day.

The $\mu$709 introduced the next major change in op amp design. It was fabricated and issued in 1965 by Fairchild Semiconductor. As compared to the $\mu$A702, it had higher gain, a larger range, lower input current, and a more acceptable power supply requirement ($\pm$15 vdc).

Further development of the LM101 op amp by National Semiconductor in 1967 provided yet more versatility. The LM101 increased the gain (up to 160 K) and the operation range. The LM101 also provided short-circuit protection. Probably the most important contribution of the LM101 is simplified frequency compensation. This was accomplished by placing an external capacitor across selected connection pins. The first op amp to provide this internally was the hybrid LH101, which was an LM101 and a capacitor in a single package.

Fairchild Semiconductor made an internally compensated device in May 1968 with the introduction of the $\mu$A741. The differences between the LM101 and the $\mu$A741 are slight. The frequency compensation was accomplished using an on chip capacitor. Offset null is accomplished by adjustment of currents in input stage emitters. On the LM101, offset is accomplished by adjusting current in input stage collectors.

In December 1968, an improved version of the LM101, the LM101A, was devised. This device provided better input control over temperature and lower offset currents.

National Semiconductor introduced the LM107, which had the frequency compensation capacitor built into the silicon chip. The LM107 came out at the same time as the LM101A.

In 1968, Fairchild Semiconductor issued the $\mu$A748. The device had essentially the same performance characteristics as the $\mu$A741. The difference was external frequency compensation.

The first multiple op amp device was Raytheon Semiconductor's RC4558 in 1974. Characteristics of the RC4558 are similar to the $\mu$A741 except that the latter uses npn input transistors. Later in 1974, the LM324 quad op amp became public. This is a National Semiconductor device. It is similar in characteristics to the $\mu$A741 in speed and input current. The LM324 is especially useful for low-power consumption. The most useful attributes of the LM324, according to some engineers, is its single-power-supply requirement.

The first FET input op amp was the RCA CA3130. With this addition to the op amp family, extremely low input currents were achieved. Its power can be supplied by a 5- to 15-vdc single supply system.

In July of 1975, National Semiconductor came out with the LF355. This was the first device created using ion implantation in an op amp.

Texas Instruments introduced the TL084 op amp in October 1976. It is a quad JFET input op amp; it also is an ion-implant JFET. Low bias current and high speed are two of its attributes.

Since 1976, the types of op amps have increased almost daily. Today we have a variety of op amps that will provide the user essentially anything he needs, such as high common-mode rejection, low-input current frequency compensation, and short-circuit protection. The designer simply expresses a need and is supplied with the correct type.

# 2
# Integrated-Circuit Hierarchy

In this chapter we will deal with some of the names and acronyms associated with integrated circuits. Throughout the electronics industry, each new discovery or situation brings forth new words, new meanings of old words, and acronyms. At first glance they may appear mysterious or confusing. We will attempt to sort some of these words out for the reader and put them into meaningful order.

Integrated circuits are solid-state devices. Solid-state devices are continuous in structure and thus have no moving parts, filaments, or vacuum gaps. An *integrated circuit* is a collection of components connected together in a unique and complete configuration to perform some useful electronic function. An integrated circuit, or IC, is a circuit whose components are manufactured on a continuous or single piece of semiconductor material. If all the components are on a single piece of material (or chip), the integrated circuit is called *monolithic*, which literally means one stone. If two or more interconnected chips or components comprise the circuit, the IC is said to be a hybrid.

In figure 2-1, the order of ICs by fabrication method is presented. The monolithic fabrication is accomplished by the planar process. The *planar process* is the basic manufacturing process of solid-state technology for both bipolar- and unipolar-function integrated circuits. It derives its name from the fact that all integrated-circuit components are interconnected on the same surface or plane of the semiconductor material. This process embodies four key principles: (1) diffusion, (2) oxidation, (3) selective oxide removal, and (4) epitaxy. This process is defined in detail in a later chapter.

Bipolar monolithic ICs are those devices whose components are current-controlled and require both positive (+) and negative (−) polarity for operation of their active elements.

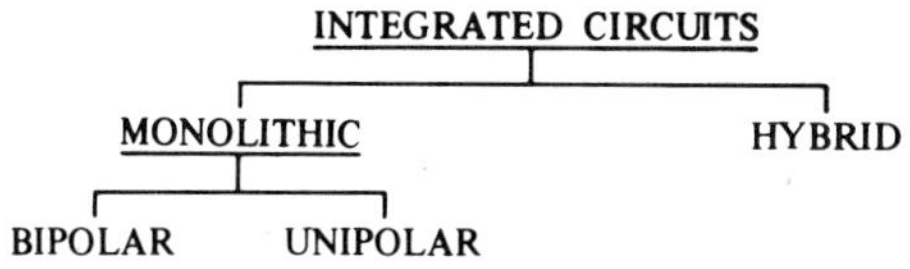


Figure 2-1 Classification of Integrated Circuits by Fabrication

Unipolar monolithic ICs are those devices whose components are voltage-controlled and have a single polarity operational current in their active elements. These devices are also manufactured by means of the planar process with variations.

*Hybrid* ICs are several devices and circuits placed on a common substrate, then connected. Other elements such as diodes and capacitors are attached to make the circuit complete. In this manner they can perform functions together. Two methods of manufacture are, thin- and thick-film.

The second comparison of integrated circuits is by function (see figure 2-2). Integrated circuits are divided into two basic function categories, digital and linear. *Digital* ICs are those whose inputs and outputs have two discrete states of operation, cutoff and saturation. *Linear* ICs are those whose inputs and outputs are proportionally or mathematically related and may vary between cutoff and saturation.

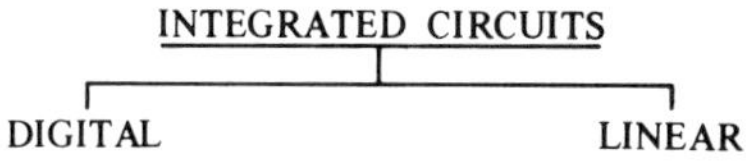


Figure 2-2 Classification of Integrated Circuits by Function

There are two other related functions of ICs, analog and nonlinear. Analog is often taken to mean linear in the integrated-circuit world. In fact, many electronics engineers and technicians use the words "linear" and "analog" synonymously.

*Analog* circuits can be related to digital and linear ICs. Analog circuits perform a variety of instrumentation, computation, and control functions. They are computational circuits. *Nonlinear* circuits may do analog computing. Nonlinear, of course, means functions that are not linear. Nonlinear circuits, however, can control a function (primarily computational). They translate a mathematical relationship between independent variables at the input and a dependent variable at the output of circuits.

## INTEGRATED-CIRCUIT COMPONENTS (See figure 2-3)

**Bipolar Components** The transistor is the major discrete-circuit bipolar component. In integrated circuits, bipolar devices are diodes, transistors, resistors, and capacitors.

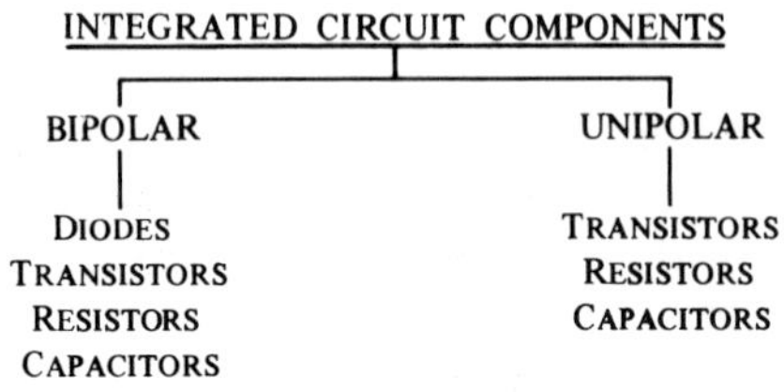


Figure 2-3 Integrated-Circuit Components

**Unipolar Components** The metal oxide semiconductor (MOS) and the junction field effect transistor (JFET) are the discrete-circuit unipolar devices. In integrated circuits the unipolar devices are transistors, resistors, and capacitors.

## DIGITAL INTEGRATED CIRCUITS

**Digital ICs by Transistor Type** (See figure 2-4) Digital integrated circuits are manufactured by the bipolar and the unipolar methods but have different electrical functions. Bipolar digital ICs utilize NPN and PNP transistors. Unipolar digital ICs utilize both metal oxide semiconductors (MOS) and field effect transistors (JFET). There are three MOS-type transistors: N diffused MOS (NMOS), P diffused MOS (PMOS), and N and P diffused MOS (CMOS). The C in CMOS is taken from the word *complementary*, meaning the N and P diffusions are both used and complement one another in CMOS operation.

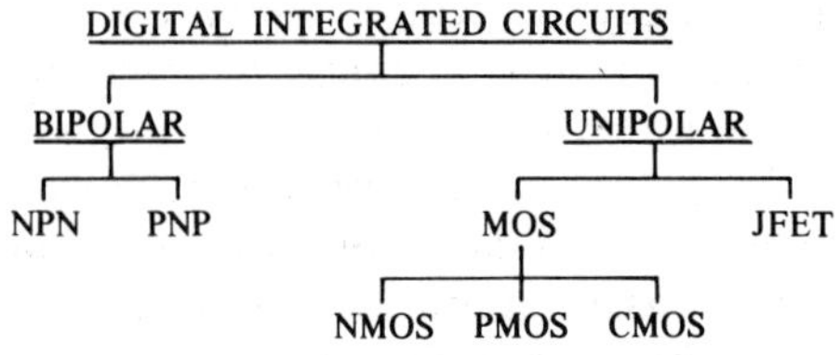


Figure 2-4 Digital Integrated Circuit by Transistor Type

**Digital Bipolar Device Complexity** (See figure 2-5) Bipolar components are classified by their complexity. The least complex is SSI or small-scale integration. These devices have up to 10 gates in their circuitry. The SSI devices are constructed to house gates and flip-flops. The second level of complexity is the MSI, or medium-scale integration. MSI devices have counters, decoders, registers, ROMs, and RAMs as their circuits. MSI devices have 10 to 100 gate circuits. LSI devices have the most gate circuits (over 100). Bipolar devices, until recently, did not compete too well with LSI devices. A new bipolar technique called $I^2L$ (integrated-injection logic) is bringing LSI bipolar devices up to the level of their MOS (unipolar) counterparts. Components within bipolar devices are made with the transistor types NPN and PNP, in all the complexity types.

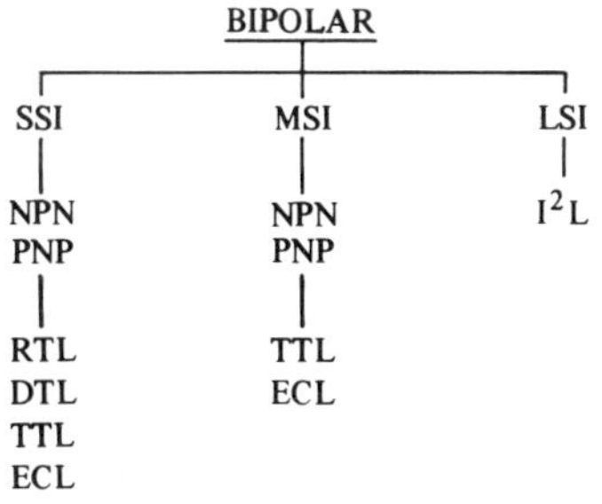


Figure 2-5 Bipolar Device Levels of Complexity

**Digital Bipolar Device Logic Families** (See figure 2-5) Most devices can be classified as part of a family of circuitry used to form logic gates. The first of these families was the diode logic. Diode logic goes back to the tube circuits of yesteryear. This circuit had diode inputs and a load resistance output. If the signal was of the correct polarity, the diode would conduct and current would flow to the load.

Probably the first of the integrated-circuit families was resistor transistor logic (RTL). The inputs were placed on resistors and the outputs were taken from transistor collectors. The RTL was low-speed and noise-sensitive. However, it was a direct translation from discrete component circuits.

The RTL family was speeded up by placing capacitors across the input resistors. The outcome was called resistor, capacitor, transistor logic, or RCTL.

The next family of logic that came into focus was the diode transistor logic (*DTL*). Diodes were used at the input to steer the signals and to speed the operation. DTL is faster and less susceptible to noise than RTL.

*TTL*, or transistor transistor logic, devices are high-speed devices. Inputs are placed on a dual (or more) emitter transistor and outputs are taken from transistors. The TTL devices generate noise and use more power than RTL and DTL devices.

The fastest device used in the logic family is *emitter-coupled logic* (ECL). The families DTL and TTL operate at two levels, off or on (cutoff or saturation). The ECL is high-speed because it operates between cutoff and saturation. The circuits have a voltage divider at the base of the logic transistors similar to a biased amplifier.

*Integration-injection logic* ($I^2L$) has low power requirement and operates at medium speed. The inputs are transistors and the outputs are transistors. There are no passive components such as resistors in the circuitry. Fabrication therefore takes fewer steps and can be much more dense than other family logic.

Digital Unipolar Device Complexity (See figure 2-6) As in bipolar devices, the complexity of this device falls into three major areas: small-scale integration (*SSI*), where there are up to 10 unipolar devices; medium-scale integration (*MSI*), with 10 to 100 unipolar devices; and large-scale integration (*LSI*), with over 100 unipolar devices.

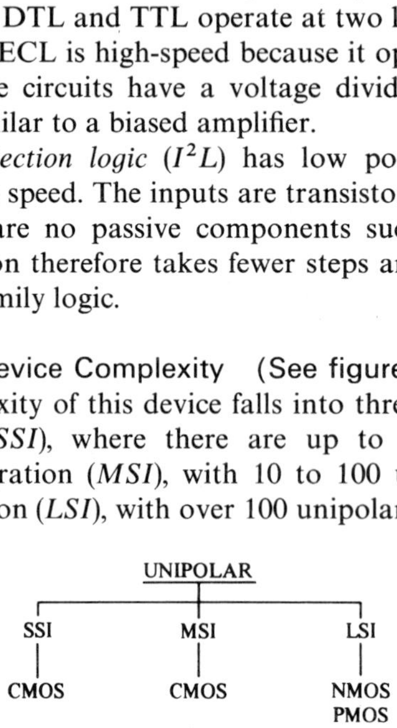


Figure 2-6 Digital Integrated Unipolar (MOS) Circuits-Levels of Complexity

Most components within the unipolar devices are made with the transistor types CMOS, NMOS, and PMOS.

Families within the unipolar devices are similar to the bipolar families.

MOS devices are noted for their low power consumption, small die size, and high noise immunity. CMOS is not easy to interface with other logic families and often impossible to match.

## LINEAR INTEGRATED CIRCUITS

Linear ICs by Transistor Type (See figure 2-7) Linear ICs have evolved from the use of transistors and MOS devices in separate devices

LINEAR CIRCUITS

- UNIPOLAR
- BIPOLAR
- COMBINATIONS

Figure 2-7 Linear Circuits by Transistor Type

to combinations that include both these fabrication techniques. The usual technique in combinations is to use a *FET* (field effect transistor) at the input. This supplies the device with an extremely low input current, a notable attribute.

**Linear Integrated Circuits (See figure 2-8)** By far the predominant linear integrated circuit is the operational amplifier. In fact, this device is used so often in the electronics industry that many books, including this one, have been written about it.

There is, as you can see by the listing in figure 2-8, a fairly large variety of linear integrated circuits. Linear integrated circuits perform process and control operations, signal generation, and amplification. The major use, of course, is with operational amplifiers. However, the voltage regulators and comparators are in increasing use.

LINEAR INTEGRATED CIRCUITS

- OPERATIONAL AMPLIFIERS
- VOLTAGE REGULATORS
- VOLTAGE REFERENCES
- INSTRUMENTATION AMPLIFIERS
- VOLTAGE COMPARATORS
- ANALOG SWITCHES
- SAMPLE AND HOLD CIRCUITS
- A/D AND D/A CONVERSION
- TELECOMMUNICATIONS
- AUDIO/RADIO/TELEVISION
- TRANSISTOR/DIODE ARRAYS

Figure 2-8 Linear Integrated Circuits

## NONLINEAR INTEGRATED CIRCUITS

**Nonlinear ICs Defined** A confusion may arise from the word "linear." *Linear* in electronic devices means that cause and effect are proportional for all values of input and output. Natural devices in real life are *not* linear. In fact, the linearity across a range of operations is limited. Devices can, however, be *nonlinear* in operation but still have useful life because the nonlinearity can be well defined, controllable, stable, and available at low cost. Analog relationships such as multiplication, square laws, and log ratios are just some of the examples of usefulness. Applications include modulation, power measurement, signal shaping, and correcting for the nonlinearity of measuring devices. In general, the nonlinear device may be classified according to its smoothness.

If the functions are smooth, it may be classified as continuous function. If it has one or more discontinuities or jumps, it is classified as a discontinuous nonlinear function.

## HYBRID DEVICES

The *hybrid* device is fabricated by placing devices, including ICs and discrete components, on a single substrate and connecting them electrically to perform electronic functions.

There are two basic techniques for fabrication of the hybrid: these are thin-film and thick-film. They are similar. *Thin-film* fabrication is accomplished by deposition of a thin layer of metal on a substrate. Substrates are coated and metal layers are selectively removed to form the circuit. Devices such as diodes, resistors, transistors, and capacitors are then attached to complete the circuit on the substrate.

In *thick-film* fabrication a heavy coat is produced by firing a paste deposited on a substrate along with other discrete components. The paste structures are conductors, resistors, and capacitors. The paste is usually stencil screened on a ceramic substrate. Components of the circuit all require a different paste composition. Firing sets the circuit.

Thick-film is cheaper than thin-film, but requires a larger substrate. Thin-film is ideal for high frequencies.

# 3
# The Fabrication Process

The development of integrated circuits has evolved into geometrics that are extremely small. The quality of the silicon substrates on which the integrated circuits are fabricated has inherently become more critical. Greater uniformity and decreased levels of impurities are demanded to increase device integrity and wafer yield. Fabrication of the integrated circuit must start with the best possible single-crystal silicon wafer.

This chapter deals with the process of producing these wafers and the fabrication of the integrated circuit.

## THE NUTSHELL INTEGRATED-CIRCUIT FABRICATION PROCESS

It must be noted that there are as many procedures for the fabrication of integrated circuits as there are companies that produce them. The processes are all similar, however, with only a few of the more exacting details held as proprietary.

The fabrication of the silicon integrated circuit begins with the growing of a single-crystal silicon ingot in an oven (crucible). The crystal is sliced, etched, and polished. The slices, now called wafers, are transported to device manufacturers who perform what is called the *planar process*. This process involves oxidation, selective dioxide removal, diffusion, and epitaxy. Oxidation is the formation of silicon dioxide layers on the silicon for photographic masking. Selective dioxide layers are removed with acids. Diffusion is the process of depositing dopant through dioxide windows. *Epitaxy* is the growing of one semiconductor upon another. The planar process places the circuit on the silicon wafer. The circuit pattern is duplicated many times on the wafer (depending on the size of both wafer and circuit). Individual circuits are called dies. Each die is attached to a package to hold it in place. Wires are connected from the die to the outside of the package. The package is closed for protection. The package with die and wires is now called an integrated circuit (IC).

This is a nutshell description of the fabrication process. We shall now look at the details involved.

## THE SILICON-WAFER GROWTH PROCESS

As integrated-circuit devices become more and more complex, the need for quality in the wafer increases. Purity is usually provided when the virgin polycrystal is etched clean and placed in a crucible. It is the process of crystal growth that really determines the perfection of the finished wafer and establishes the properties of the substrate dies.

Primary among the crystal growth processes is the Czochralski (CZ) method (used by 80 percent to 90 percent of the world market). The float-zone method is the only other method that controls any part of the industry. Float-zone crystal growth may be used in 10 to 15 percent of the world market for wafers. There are other methods that have been studied, including the ribbon method and the heat exchange method. In this section we shall discuss these methods. The Czochralski (CZ) method will be covered in detail, while only a cursory look will be taken at the alternates.

### The Czochralski (CZ) Crystal Growth Method (See figure 3-1)

As has been stated, the CZ method for growing silicon crystals has dominated the industry for some time. Other methods, primarily the float-zone method, are also used; but the future, according to experts in the field, is with the CZ method.

The entire process begins with one of the most abundant elements, silicon. Silicon in its natural state is simply sand. Sand is refined to produce semiconductor-quality polycrystalline silicon (99.9999 percent pure). The silicon and a small amount of dopant are placed in a crucible and melted. The silicon and dopant are called a *charge*. A seed crystal is dipped into the melt at a temperature near solidification. The seed is raised slowly so it will grow solid-phase crystal at the solid/liquid interface. The solid grown crystal assumes a shape much the same as a roll of bologna. The solidified silicon takes the crystalline orientation of the seed crystal. Diameter of the ingot is controlled by temperature and seed pulling rate. Typical diameter is 2 to 4 inches, with the length dependent on the furnace size and the melt volume.

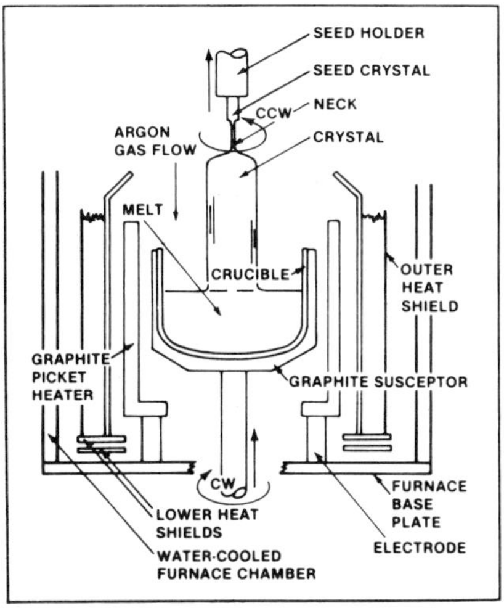


Figure 3-1 The Czochralski Crystal-Growing Process (Courtesy, Siltec Corp.)

**Polycrystalline Silicon** The charge raw material is called virgin polycrystalline silicon and is known as "poly." This material is the final base for all integrated circuits and semiconductor devices.

Silicon in its raw state is usually silicon dioxide (sand). Some sand may be found to be from 95 to 99 percent pure. The sand is mixed with carbon to form 99 percent pure silicon and carbon dioxide. The silicon is then reacted with hydrogen chloride to form trichlorosilane. Decomposition of the trichlorosilane with an electric current forms rods of virgin polycrystalline silicon (poly). The polycrystalline silicon contains many crystals. One of the purposes of the crystal growth, of course, is to enable this material to be converted into a single-crystal configuration.

Even with the finest production methods, pure silicon does not exist. The presence of infinitesimal amounts of elements such as boron and phosphorous cannot be avoided. The amounts are being held in the range of tenths of a part per billion.

**The Dopant** As you may know, purified silicon (a semiconductor material) has exactly four electrons in its valence. The atomic structure of silicon lies between good conductors such as copper and silver and good insulators such as glass or ceramic. The dopant element is one of two types, N or P.

The N-type element (also called a donor atom) has five electrons in its valence. Figure 3-2A shows a model of silicon with an atom of N-type material (phosphorous) in its crystal lattice structure. The reader will note that the donor atom has a loosely bound electron. Typical donor elements (atoms) are phosphorous, arsenic, and antimony.

The P-type element (also called an acceptor atom) has three electrons in its valence. Figure 3-2B shows a model of silicon with an atom of P-type material (boron) in its crystal lattice structure. Note that the acceptor atom has a space called a *hole* (absence of an electron). Typical donor elements (atoms) are boron, aluminum, and gallium.

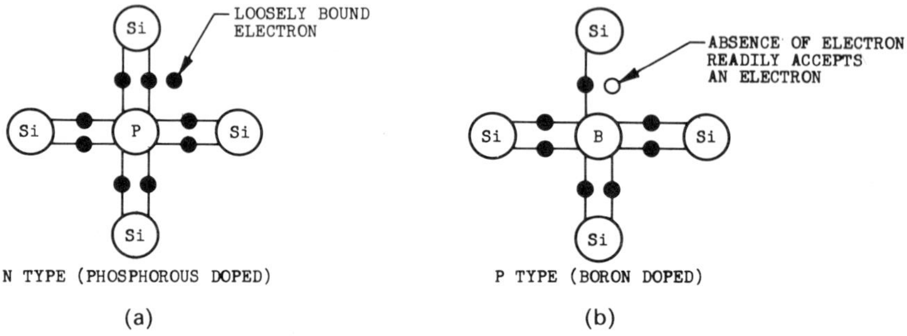


Figure 3-2 Dopant Elements (Courtesy, Siltec Corp.)

N-type material is silicon that has been doped to be electrically conductive by means of mobile electrons and therefore is negative. Conversely, if the silicon has been doped for conduction by holes, it is said to be positive or P type. The type of dopant (N or P) is extremely important to determine bias polarity in the design of electrical circuits.

The second important characteristic of the dopant is resistivity. A semiconductor is, as its name implies, semi- or partially conductive. The degree of conductivity of the silicon is fixed by the amount of dopant that is added. The resistivity of the end product (semiconductor device) is determined by the type and amount of the dopant material added to the silicon. More will be said about resistivity when we discuss the wafer and its fabrication.

**The Charge** Ultrapure polycrystalline material (Si) is etched clean and weighed to ensure that the charge is the correct size. The *charge* is the correct amount of material required to draw a crystal ingot of a specified size. At the same time, dopant is carefully weighed and added to the polycrystalline material. The amount and type of dopant is determined by specification and customer requirement. Additions of the dopant element are usually in the range of 2 to 50 parts per billion (ppb) or a few hundred micrograms for the typical charge size and resistivity range. Typical charge sizes are 10 kilograms (kg) for 3-inch-diameter crystals to 30 kg for 100-millimeter (mm)-diameter crystals.

**The Melt** The charge including the polycrystalline silicon (Si) and the dopant is placed into a previously cleaned quartz crucible. (See figure 3-1.) The crucible sits in a carbon susceptor. The susceptor and the crucible are heated by either the induction (RF) or thermal resistance method. The susceptor and crucible are rotated by some external drive. The furnace components—such as heat shields, top plates, and plumbing fixtures—are put in place. The sealed furnace chamber is filled with an inert gas (usually argon). The entire furnace is checked for leaks. The meltdown then begins; this takes from one to three hours, depending on the size of the charge. Constant monitoring of the charge is required to prevent the polycrystalline material from sticking to the walls of the crucible. Following meltdown, the melt is stabilized at a temperature between 1415° and 1420° Centigrade (C) and the crystal growth may begin.

**Crystal Growing** The term "crystal" refers to a material that is highly ordered on a long-range atomic basis. The general process of converting polycrystalline silicon to single-crystal silicon is called *crystal growing*. The conversion is from a random to a symmetrical crystalline structure. More information on the crystalline structure will be provided when the wafer is discussed later in this chapter.

The process begins. A CCW rotating seed crystal of the desired crystal orientation is lowered by an operator and dipped into the CW rotating melt. The temperature of the melt is near the solidification point. If all conditions are correct, the lifting of the seed crystal will cause the transition to solid-phase crystal at the solid/liquid interface. Once this occurs, several operations must follow in sequence: necking, shouldering, body growth, and roundoff.

*Necking* occurs by growing a small-diameter section from the seed at high pulling speeds. Necking is required to prevent defects in the crystal called dislocations. Dislocations are imperfections in the crystalline structure caused by missing or unbonded atoms in the lattice. Dislocations must be eliminated by the operator, who identifies proper growth patterns by subtle changes in the surface appearance of the neck.

*Shouldering* is done by slowing the pull speed and adjusting the temperature to allow the neck to flare out until 90 percent of the growth diameter is attained.

At this point, *body growth* is set with adjustments to rotation, pull, and temperature to obtain a cylinder of the desired diameter. The body growth continues until the bulk of the melt is grown.

During growth operations, the dopant distribution varies through the length of the ingot. N-type crystal resistivity range may be 2.5 to 3.5 greater at the top of the ingot as compared to the bottom. P-type crystal resistivity range may be 1.7 to 2.2 ratio. This ratio, called segregation coefficient, is critical for usable crystal yields.

*Roundoff* is achieved at this point by increasing pull speed, thereby decreasing diameter, until there is only one point of contact with the melt left. The crystal is then cooled, removed from the furnace, and evaluated for physical and electrical parameters. Cooling takes one to two hours. Heat treating in a separate furnace may be required to stabilize electrically active oxygen.

It must be noted that operator skill is crucial in the growth process. Each step is vital to achievement of a perfect crystal ingot. Seed dipping, necking, shoulder creation, body growth, and roundoff all require expertise that is derived by training and experience.

**The Float-Zone (FZ) Crystal Growth Method** The first alternate method used for growing single-crystal ingots is the float-zone or FZ method. It runs a distant second in produced volume to the Czochralski (CZ) method just discussed. The FZ method is much faster than the CZ method, but costs more because of small-diameter ingots. FZ is still required for high-power rectifiers and thyristors where high resistivity is necessary. Some FZ crystals are used with integrated circuits.

The FZ method operates in the following manner. A polycrystalline rod of silicon is placed in a chamber. The chamber is evacuated or filled with argon gas. The top of the silicon rod is attached to a chuck. A silicon seed crystal is attached to a second chuck at the base of the chamber, with its center concentric with the center of polycrystalline rod. An RF heating coil is placed around the rod. To begin the growth, the RF heating coil forms a molten drop at the bottom of the rod. The seed crystal is brought up to the molten drop. Controlled movement of the rod and the seed through the RF coil tapers the rod, then gradually develops the correct diameter. The growth may be in either direction, but is usually up the rod from bottom to top. The molten growth zone is around 2 centimeters in length. Shoulder taper, body growth, and roundoff are similar to the CZ method. The finished ingot is the same as with the CZ method.

**Other Crystal Growth Methods** There are, of course, other crystal growth methods. Over the years, as we have mentioned, the Czochralski and float-zone methods seem to predominate. In fact, according to the professionals in this business, the CZ method will probably continue to dominate the field—that is, at least for larger-diameter ingots.

The *ribbon growth method* has been used to grow sapphire crystals. The method is called edge-defined growth (EFG). Several ribbons of 2-inch-wide sapphire crystal are pulled at one time from one crucible. This method eliminates the kerf loss (sawdust) that results when the ingots are sawed into slices/wafers.

Another method is the *heat exchanger* or Schmid-Vichnicki. This method is used with silicon. The polycrystalline silicon is melted in a crucible. A small solid seed is kept in the bottom using a heat exchanger. Heat is removed by strict control that allows the melt to crystallize from the seed.

## SOME CHARACTERISTICS OF THE SILICON WAFER

Before we go any further, it seems appropriate to provide some information on peculiar characteristics of the silicon wafer. These are important, in that the processes that follow have been developed so as to adhere to the requirements of these characteristics.

**Resistivity (See figure 3-3)** The resistivity of a material is the degree with which it will or will not conduct electricity. The resistance of copper is low because it is a good conductor. The resistance of glass is very high because it is a good insulator. Materials that are not great conductors but also not poor conductors of electricity are semiconductors. Silicon is one of the best and most often used electrical semiconductors. The degree with which the silicon will conduct electricity is fixed by adding dopant. Resistance, as you know, is a common electrical unit of measure.

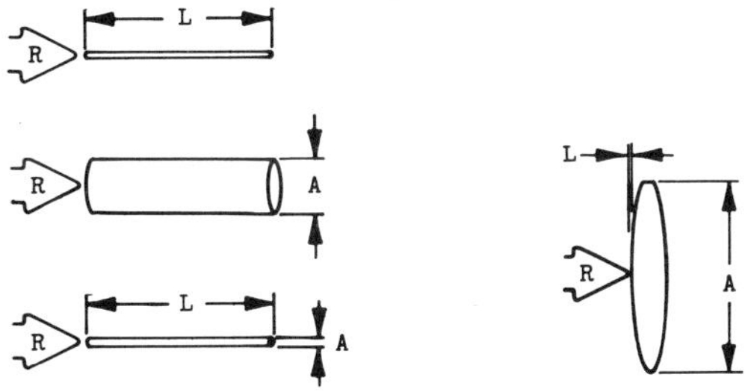


Figure 3-3 Resistivity (Courtesy, Siltec Corp.)

Consider the illustration in figure 3-3. In the top view it is readily seen that resistance ($R$) of this material is proportional to its length ($L$) in centimeters:

$$R \text{ is proportional to } L$$

In the next view down, the resistance ($R$) varies in inverse proportion to its cross-sectional area ($A$). $A$ represents the area (in $cm^2$), whether it be a rectangle, sphere, or whatever:

$$R \text{ is proportional to } \frac{1}{A}$$

If we combine these two proportions, we get a third proportion, where resistance is proportional to the length and inversely proportional to the cross section:

$$R \text{ is proportional to } \frac{L}{A}$$

This proportion is valid for a piece of wire a mile long or a small half-watt resistor, or indeed, a wafer (see figure 3-3).

The resistivity of a silicon wafer is designated by the Greek letter $\rho$. A unit of resistance is designated by the Greek letter $\Omega$. $\rho$ is dependent on how much dopant that is added to the silicon during the growth process; it is defined as the specific resistance of a material.

If we introduce the specific resistivity $\rho$ into the proportion just discussed, we have a more exact equation:

$$R = \frac{\rho L}{A}$$

or

$$\rho = \frac{RA}{L}$$

where

$R$ = resistance in $\Omega$

$\rho$ = resistivity in $\Omega$ per cm

$A$ = cross-sectional area

$L$ = length

The unit we are dealing with in this section is the silicon wafer. The wafer is a slice or a sheet of silicon. Measuring sheet resistance ($R_S$) of the silicon wafer is accomplished using one of two methods—the contact method or the noncontact method.

**The Contact Method of Measuring Resistivity of a Wafer** With the contact method, a four pointed probe is utilized. This probe is precision made. The probe points are aligned on the silicon in linear array and equally spaced. (The most common spacings used in industry are 0.025 inch and 0.0625 inch, with the latter by far the most popular.) A current is passed through the outer two probe points and through the wafer under measure. The voltage drop is measured across the center two probe points. Resistivity is then calculated from the magnitude of the current applied and the voltage drop measured. The formula that relates resistivity to current and voltage is as follows:

$$R_S = 4.53 \frac{E}{I}$$

where $R_S$ = sheet resistance

$E$ = voltage

$I$ = current

This equation is useful where the sheet thickness is very much less than the probe spacing and the sheet length is very much greater than the probe spacing.

Once the sheet resistance is known, the resistivity is calculated by the following formula:

$$\rho = R_S t$$

where $\rho$ = resistivity

$R_S$ = sheet resistance

$t$ = sheet thickness

If the wafer thickness is greater than the probe spacing, the following equation is used for resistivity calculations:

$$\rho = 2\pi S \frac{E}{I}$$

where $\rho$ = resistivity

$\pi = 3.1416$

$S$ = probe spacing

$E$ = voltage measured

$I$ = current applied

Of course, if the material is pure and concentric, curves may be drawn which relate the values just calculated.

The contact method is a National Bureau of Standards (NBS) method and recognized throughout industry. It is easy and accurate to use. It is, however, subject to inaccuracies owing to probe-tip wear and surface variations.

**The Noncontact Method of Measuring Resistivity of a Wafer** The second accepted method of measuring wafer resistivity is the noncontact method. Here, an RF oscillator generates a magnetic field between two conductance sensors. When a wafer is placed between the sensors, an eddy current is induced in the wafer which dissipates power from the RF oscillator. The power dissipated is proportional to the wafer conductance.

**Radial Gradient** When the four-point probes are placed on the silicon wafer, the resistivity may vary between the center and the edge of the wafer. This variation is specified as a tolerance between the center reading and the half-radius point. Figure 3-4 illustrates the radial gradient.

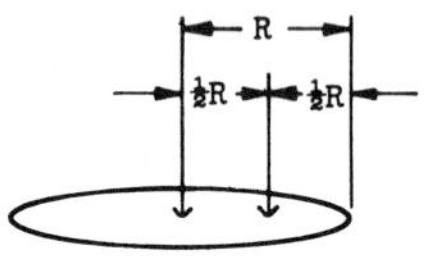


Figure 3-4 Radial Gradient (Courtesy, Siltec Corp.)

Radial gradient occurs from the pulling process. Although the dopant atoms are uniformly mixed in the melt, the phenomenon of radial gradient occurs because of the concave form of the liquid/solid interface. As the outside of the ingot in the crucible is cooled with the argon gas to facilitate growth, it freezes to solid before the center, therefore forming a concave interface shape.

**Temperature-Correction Factors** The resistivity of silicon is sensitive to temperature. The effect of temperature may be as much as 1 percent for each 2°C of deviation from 23°C standard temperature. If the resistivity measurement system used is not automatically compensated for temperature changes, then correction factors must be used. (Refer to the wafer manufacturer for these correction factors.)

**Wafer Thickness** The practice of using nominal thickness instead of actual thickness for resistivity calculations is a popular time-saving procedure. However when nominal thickness is used, errors are often introduced. If actual wafer thickness differs by 1 mil, the resistivity error will be about 4 percent for 25-mil wafers and 7 percent for 15-mil wafers.

Physical Characteristics of the Wafer Each wafer must adhere to strict physical characteristics: orientation of crystal, flatness, thickness, taper, and bow.

**Crystal Orientation** (See figure 3-5) The most prevalent and convenient method for crystallographers to define the plane in a crystalline structure is by means of the Miller indices. The indices are determined by the point at which the plane intersects the crystal axes (*x*, *y*, and *z*). For a cubic structure such as silicon, the Miller indices designate three different

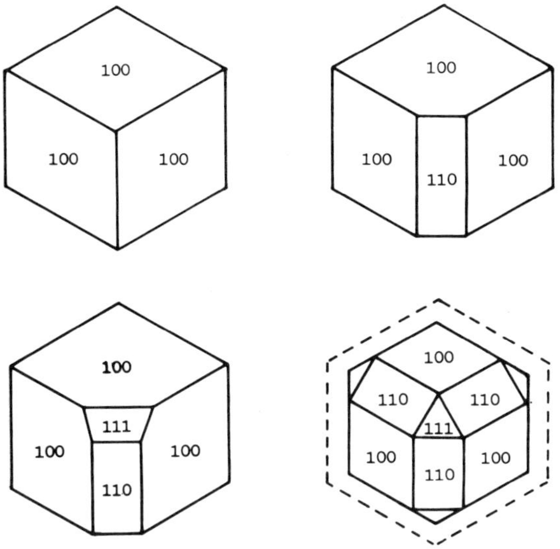


Figure 3-5 Crystal Orientation (Courtesy, Siltec Corp.)

planes. These are 100, 110, and 111. For our purposes here, the planes shown are all symmetrical. The illustration simply differentiates between the three basic planes. In actual practice, silicon cannot be grown on the 110 plane. It is only available in the 100 and 111 configurations.

The 100 orientation intersects the *x* axis so that rectangles are formed. When broken or shattered, the crystal will form rectangular pieces. The 111 orientation intersects all three axes (*x*, *y*, and *z*) so that triangles are formed. When broken or shattered, the crystal will form triangles.

**Off Orientation/On Orientation** The ingot is grown in an on orientation, and off orientation is accomplished during the slicing process. In the upper view of figure 3-6, a wafer in the on orientation is shown. This view is representative of the 100 orientation slices. In the lower view, a wafer is sliced in off orientation. It is normally desirable for 111-orientation wafers to be cut in this manner because this enhances epitaxial growth in later processes.

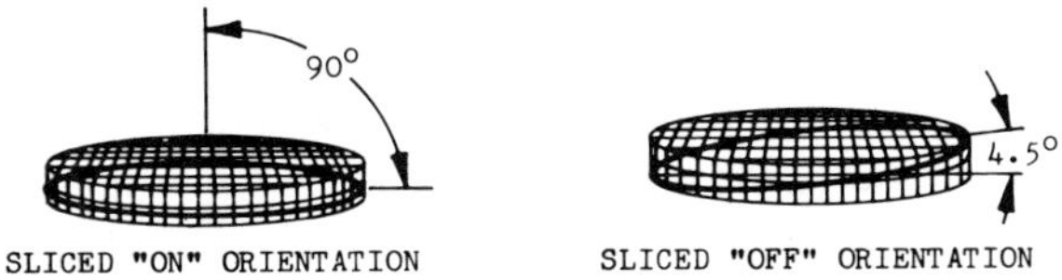


Figure 3-6 On/Off Wafer Orientation (Courtesy, Siltec Corp.)

**Flat Orientation** In figure 3-7, a wafer is shown with a flat cut across its top side. The flat is as we would look at the surface of the wafer. Flat orientation is always specified as parallel to the $1\bar{1}0$ plane. In the illustration, the flat is parallel to a natural cleavage plane. This flat indexes the wafers so that they may be scribed and broken after the devices are made. The flat is necessary for both 100 and 111 orientations. Some manufacturers use two flats for identification purposes.

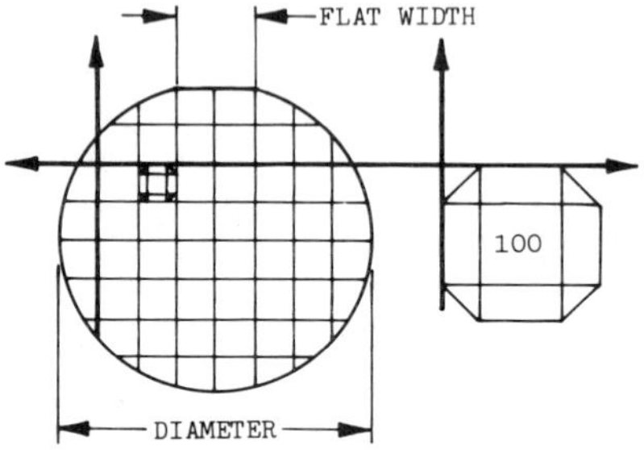


Figure 3-7 Wafer Flat Orientation (Courtesy, Siltec Corp.)

**Diameter** The wafer diameter is specified by the customer. This diameter will include a tolerance. Out-of-round diameters may also be specified. In the case of the off-orientation crystals, a slightly elliptical wafer will result.

**Thickness, Taper, Bow** These three parameters are shown in figure 3-8. All are independent of each other. Note that thickness is always measured at the center of the wafer.

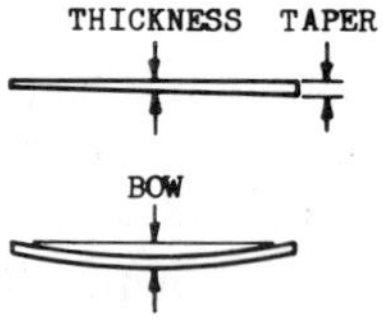


Figure 3-8 Wafer Thickness, Taper, and Bow (Courtesy, Siltec Corp.)

**Waves and Dimples** (See figure 3-9) These are specified as an allowable amount of distortion on the wafer surface. If a grid is reflected on the surface of a wafer, then any surface variation will appear as a lateral distortion of the grid. The amount is specified as a deviation.

Visual Defects (See figure 3-10) Visual defects concern the way the wafer looks; they determine the cosmetic value of the wafer.

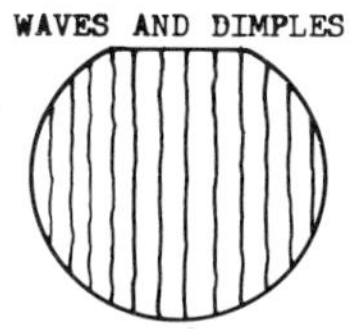


Figure 3-9 Wafer Waves and Dimples (Courtesy, Siltec Corp.)

**Scratch** Figure 3-10A shows a scratch. Macroscratches are those scratches that can be seen with the unaided eye using normal light. Macroscratches are observed using a high-intensity light at an oblique angle in a dark environment.

**Chip** Figure 3-10B shows a chip. A chip is a defect at the edge of the wafer that goes all the way through the wafer. Deep V-type chips are especially undesirable.

**Flake Chip** Figure 3-10C shows a flake chip. Similar to the regular chip, the flake chip begins on the edge but only affects one side of the wafer.

**Crack** The fracture shown in figure 3-10D starts at the edge of the wafer and moves toward the center.

**Crowsfoot** A crowsfoot, shown in figure 3-10E, is a fracture in the body of the wafer and follows the crystalline orientation. It generally appears as several intersecting cracks.

**Pit** The pit in figure 3-10F is a microscopic depression in the polished surface of the wafer.

**Dirt** Dirt, as shown in figure 3-10G, is any contaminant that can be seen with the unaided eye using high-intensity oblique light.

**Orange Peel** (See figure 3-10H) This is a roughness characterized by multiple rounded irregularities.

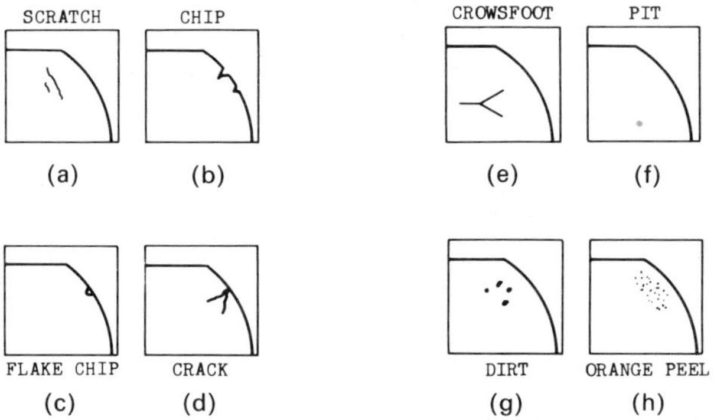


Figure 3-10 Wafer: Visual Defects (Courtesy, Siltec Corp.)

**Intrinsic Characteristics** These are, of course those characteristics that are developed in the crystal growth process. They are not caused by handling, but are inherent in the manufacturing process. A major part of the engineering task is to prevent these from happening.

**Dislocations** Dislocations are imperfections in the crystalline structure, caused by missing or unbonded silicon atoms in the lattice. They may be introduced in the growing process, and may also be caused by work damage during slicing, grinding, or polishing the wafers. Dislocations are observable after a Sirtl etch by microscopic analysis. This preferentially etches sharp areas, resulting in microscopically observable pits.

**Blotch Haze** This is the name given to a milky pattern observed after initial oxidation and etching. The effect is caused by shallow etch pits, caused by the preferential etching of minute imperfections which attract impurities.

Stray metal atoms may cause leakage problems, refresh problems, or softening of the diode characteristics of bipolar and MOS devices, and are generally undesirable.

It is almost impossible to avoid the introduction of trace amounts of metal contaminants at the wafer surface when manufacturing silicon wafers. A preferred method for eliminating the problem is to introduce a controlled amount of damage on the back side of the wafer. The damage creates high-energy sites at the back of the wafer, which attract and trap the metal atoms during the elevated temperature of oxidation.

**Slip** (See figure 3-11A) Slip is a plastic deformation where one part of the crystal shears, relative to another part. The displacement takes place along a definite crystallographic direction in the wafer.

**Swirl** (See figure 3-11B) Swirl can be detected after initial oxidation and Sirtl etch, and appears in the form of concentric rings on the polished surface of the wafer. Microscopically, these are point defects caused by vacancy clusters, due to vacancy-oxygen complexes acting as nucleation sites for a supersaturated amount of vacancies. Oxygen, carbon, and heavy metals can precipitate at these point defects, causing soft junctions and leakage problems.

Figure 3-11 Wafer: Intrinsic Characteristics (Courtesy, Siltec Corp.)

**Wafer Flatness** The trend toward smaller device geometries to allow more dies per wafer causes a demand for wafer flatness. It is difficult to compensate for lack of flatness at the optical level. If the wafer is not flat, the image (circuit parts) will be distorted.

There are currently three terms used in the wafer world to describe flatness. The first is "total indicated reading" (TIR) or peak to valley (PV). In figure 3-12A, this is shown as the difference in elevation between the highest and the lowest points on the wafer.

"Nonlinear thickness variation" (NTV), shown in figure 3-12B, describes the thickness variation of a wafer defined by the center thickness value and four edge thickness values obtained $\frac{1}{8}$ inch from the edge.

"Focal plane deviation" (FPD), shown in figure 3-12C, is the peak deviation + or − of the wafer surface from a reference plane. The reference plane is a focal plane established in projection printing equipment, by three points on the wafer periphery.

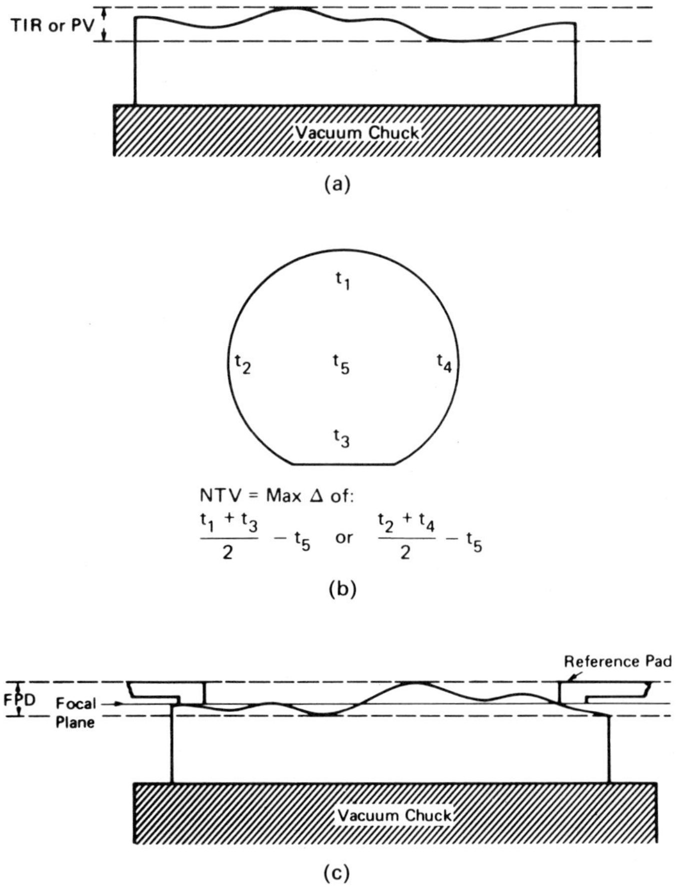


Figure 3-12 Wafer Flatness (Courtesy, Siltec Corp.)

## A WAFER MANUFACTURING PROCESS

Some of the methods for growing wafer crystals, along with the characteristics of the wafer, have been discussed. This section deals with the details involved in the actual manufacturing process from polycrystalline silicon to the finished wafer that is shipped to integrated-circuit manufacturers. The procedure (less proprietary information) is presented in the next several paragraphs. The photographs in figure 3-13 were provided by the Siltec Corporation, Menlo Park, California.

**The Process** The wafer manufacturing process begins with virgin polycrystalline silicon (figure 3-13A). The polycrystalline silicon is etched clean and placed in a quartz crucible. Dopant is added to achieve a precise electrical resistivity. Dopant is chosen to provide N- or P-type characteristics. In a specially designed crystal-growing furnace (figure 3-13B), the crucible with its doped silicon is heated to melting temperature. A small seed crystal with the desired growth orientation is dipped into the melt (figure 3-13C). The growth process begins. The seed and the crucible are turned simultaneously at specified speeds and in a controlled environment. As the crystal seed is slowly pulled, the molten silicon crystalizes and forms a single crystal called an ingot (figure 3-13D). (These last two photographs were taken through a window placed in the top of the growing furnace. The window is used for operator viewing of the growth process.) This particular process we are discussing is the Czochralski method. After growth is completed, the furnace is allowed to cool and the ingot is removed. A permanent ingot number is assigned in the growing operation which identifies the electrical characteristics of the ingot, therefore the wafer. The ingot is annealed to stabilize its electrical characteristics, and a resistivity profile is made of the entire ingot. The ingot is placed in a container to await the next process (figure 3-13E).

The ingot ends are cropped and the ingot is cylindrically ground to a specified diameter (figure 3-13F). A record is made of the diameter profile. Growth orientation is verified by the X-ray topography (Laue) technique and an instant photograph is made to serve as a permanent record (figure 3-13G). Using this information, a precisely oriented flat is ground on the ingot for future alignment reference by the integrated-circuit manufacturer (figure 3-13H). Vertically mounted ingots are sliced by an internal-diameter diamond saw. Slice thickness is controlled by digital stopping controls (figure 3-13I). The wafers may be specified for an "off" surface orientation, requiring the ingot to be sliced at a precise angle from its axis. After slicing, the wafers are cleaned and a second Laue (X-ray) photograph verifies slice orientation. In accordance with the ingot identification, the slices' physical and electrical characteristics

A. Polycrystalline Silicon

B. Crystal-Growing Furnace

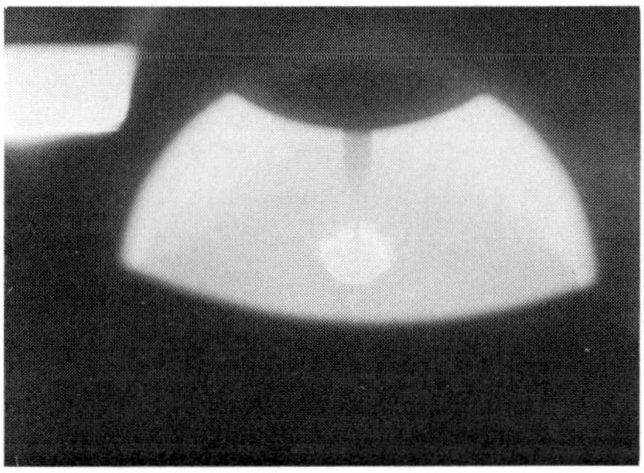

C. Seed Crystal

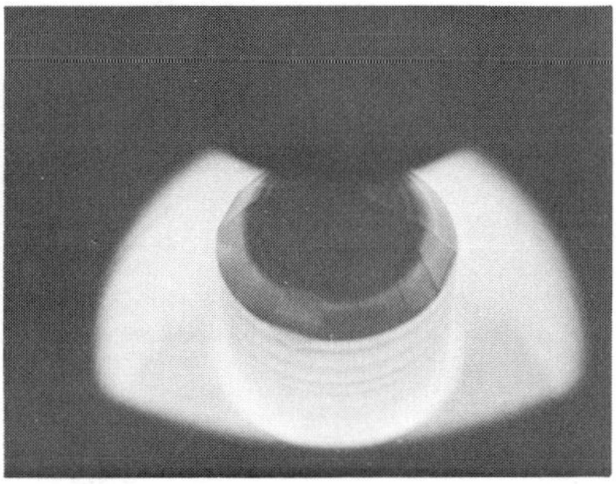

D. Forming of the Ingot

E. The Finished Ingot

F. Grinding the Ingot to Diameter

**Figure 3-13** The Wafer Fabrication Process

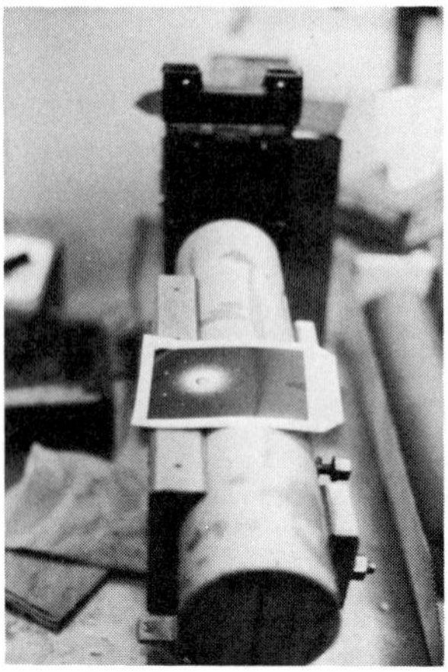

G. An X-Ray Photograph

H. The Flat

I. Slicing

J. Lapping

K. Edge Grinding

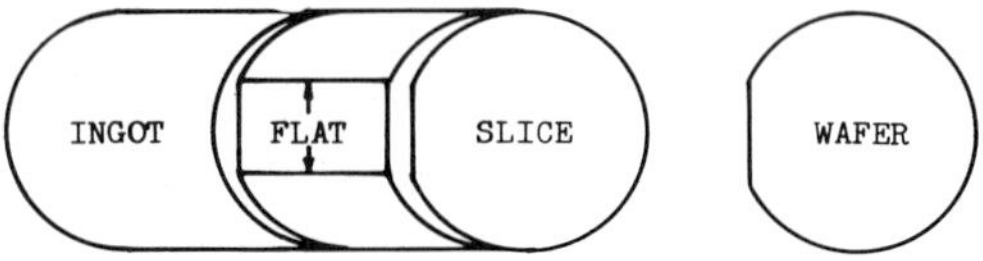


L. Ingot to Wafer

Figure 3-13 *continued*

are analyzed and recorded. The slices are then loaded into a lapping machine, where they are lapped to ensure flatness (figure 3-13J). After lapping, the slices are edge-ground; this reduces edge chips, photoresist beads, and epitaxial crowns (figure 3-13K). (The latter two problems arise in the circuit fabrication process.) The slices are then placed in an acid bath to etch (remove) any damage that may have occurred to their surfaces during the fabrication process. After etching, the slice is called a wafer. The wafer is thinner from the etching process and more like the finished product. Figure 3-13L is a resume of what has been accomplished so far—growth of the ingot, diameter grinding, flatting, slicing, and etching, along with many quality control steps.

The wafers are then checked by production inspectors for defects such as chips, scratches, discoloration, etc. Wafers not passing stringent examinations are immediately rejected. The approved wafers are placed in boats (carriers) and taken to an automatic sorting machine, where they are first tested for resistivity and thickness. The automatic sorting machine then separates them into bins according to their resistivity and thickness (figure 3-13M). If they are out of spec, they are rejected. Again the wafers are placed in plastic containers by quality inspectors to await the polishing process.

When they are called for the polishing process, the wafers are inspected for quality specifications. Then they are directed to a mount station, where they are placed onto a carrier by the operator. The loaded carriers are placed in a mounting machine, where pressure and heat are used to make sure that the wafers are firmly attached to the carriers (figure 3-13N).

The carriers, with the wafers firmly attached, are held by a vacuum system on the inside lid of the polishing machine (figure 3-13O). Polishing takes two steps; the first removes some stock from the face of the

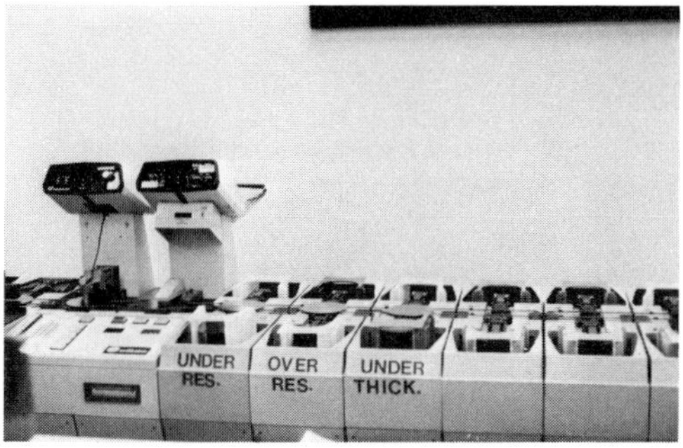


M. Automatic Sorting

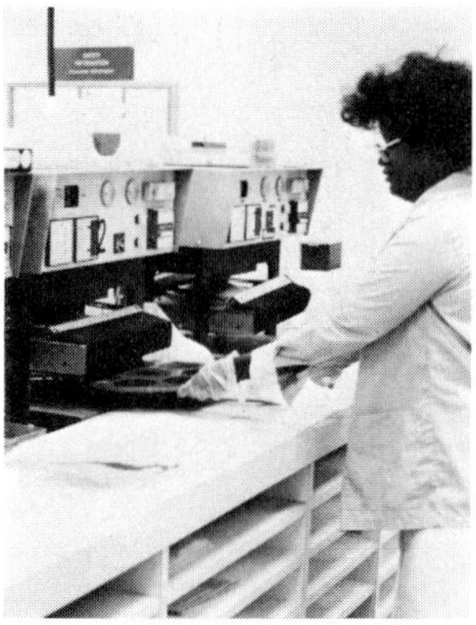

N. Attaching Wafers to Polishing Carriers

Figure 3-13 *continued*

wafer and provides an initial polish which creates a mirrorlike finish on the wafer. (Note that only one side of the wafer will be polished.) Then the carriers are removed from the finish polishers and taken to a darkened area, where the wafers are checked for damage under a high-intensity light.

The wafers are then taken to a demount station. The operator removes the wafers from the carrier by means of a vacuum (suction) wand and places them in a boat separating the wafers from one another (figure 3-13P). The boats are submerged in de-ionized water to keep the wafers free from contamination.

Detachable handles are clipped onto the boats so that the wafers can be dipped into a series of mild acid and rinse solutions (figure 3-13Q). This process removes particles or residue from the wafers' surface.

Now that all of the wafers have perfect mirror surfaces, they go into a dust-free super-clean room. A final cleaning takes place in a machine called a scrubber. They are then inspected finally and separated into lots (figure 3-13R). Then they are packed in covered containers to protect them from damage or surface contamination. Identity of resistivity and thickness are printed on the container. The container is placed in storage to await delivery to the customer.

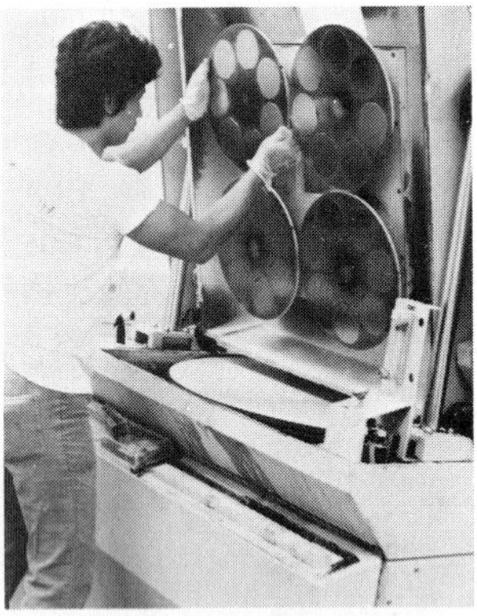

O. Connecting Polishing Carriers to Polisher

P. Removing Wafer from Polishing Carriers

Q. Cleaning and Rinsing Wafer

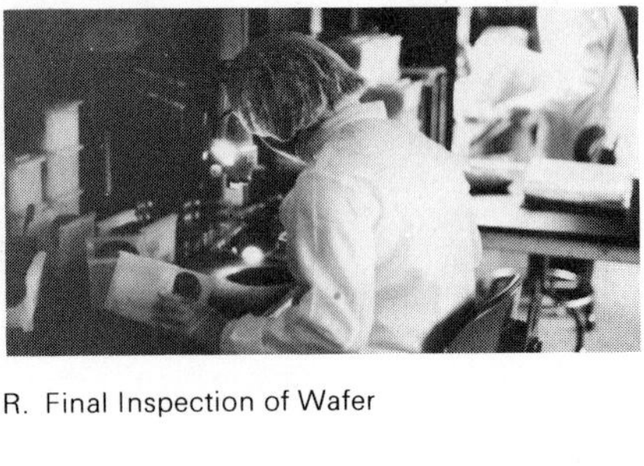

R. Final Inspection of Wafer

**Figure 3-13** *continued*

## THE FABRICATION OF INTEGRATED CIRCUITS

The integrated circuit (IC) is, of course, the final product of the fabrication process. It is an exacting and delicate process to create an IC, as you have seen from the wafer growth process just discussed.

The process must start with the purest possible wafer. The wafer must be free of defects such as discoloration, scratches, chips, and cracks. The surface should be mirror-smooth and absolutely flat. A perfect wafer is a good start to a successful fabrication process.

**The Planar Process Defined** The basic process for fabrication of all solid-state devices is the planar process. The word "planar" refers to the process of bringing all leads to the top surface of the plane of a wafer, where they can be connected or bonded. There are four major concepts involved in the planar process: oxidation, selective dioxide removal, diffusion, and epitaxy.

**Oxidation** Silicon chemically reacts with the oxygen of water vapor to form an oxide called silicon dioxide ($SiO_2$). Basically, time and temperature determine the thickness of the oxide on the wafer surface. The properties of silicon dioxide make it one of the most useful of microcircuit materials. The formation of the oxide layer is easily controlled, the layer is uniform and continuous, and it closely adheres to the silicon surface. It is also an effective barrier for the more important diffusants, but is easily removed by a chemical solvent. These two factors make $SiO_2$ ideal as a diffusion template or screen. The oxide layer is a good dielectric, finding further value in the fabrication of capacitors and passivating or protecting surfaces.

**Dioxide Removal** The photochemical procedure of removing selected areas in the oxide layer is similar to that used in the manufacturing of printed circuit boards. The wafer with the oxide layer is uniformly coated with a layer of a chemical called *photoresist*. The photoresist is dried and covered with a glass photomask that contains the desired patterns. The photoresist is then exposed to ultraviolet light through the mask. The nonopaque areas are polymerized (i.e., converted to a new compound), so they can resist attack by acids and solvents. The nonpolymerized photoresist and oxide are etched away with an acidic solvent. The wafer is now ready for diffusion, which will occur only in the openings, called windows, in the oxide. The complete photoresist process is repeated each time the silicon dioxide is selectively removed. The selective oxide removal by the photoresist process is described in detail under "photomasking" in the following pages.

**Diffusion** Solid-state diffusion is the process of implanting impurity or dopant atoms in a single-crystal structure. The doping atoms, depending on their number, are capable of modifying the electrical characteristics of the N- or P-type wafer or substrate. In diffusion, the wafers are placed in carriers or boats and loaded into a furnace through glass tubes. In the furnace, which is at about 1100°C, dopant gases are passed over the wafers. Dopant atoms from the gas diffuse into the wafers. The results of this process are dependent on (1) concentration, type, and distribution of dopant; (2) time; (3) temperature; (4) pressure; and (5) environmental conditions. Two of the common dopant sources are phosphorous oxychloride ($POCl_3$) and boron tribromide ($BBr_3$). The phosphorous in $POCl_3$ converts silicon to N type, and the boron in $BBr_3$ converts silicon to P type.

Several diffusions are made in fabricating ICs. Each one is labeled with a name related to the function or component lead with which the diffusion is associated. The isolated diffusion creates separate regions in the epitaxial N layer (on a P wafer) for each IC component. The epitaxial N layer is a nondiffused layer and forms the collector of NPN transistors and the cathodes of circuit and isolation diodes. The isolation diffusion electrically separates the various areas of the wafer containing circuit elements. The base diffusion forms the base of NPN transistors and the anodes of circuit diodes and P-type resistors. It is a shallow diffusion requiring closer process control. The emitter diffusion forms the transistors' emitters, low-value N resistors, MOS capacitor plates, and low-resistance terminals for N collector regions. A fourth diffusion, called a buried layer, is sometimes used. Its function is to form a low-resistance region under the collectors of NPN transistors. The four key diffusions are illustrated in figure 3-14.

**Epitaxy** A second, often used, technique of fabricating a P-N junction is called *epitaxy*. Epitaxy is the growing or deposition of one semiconductor upon another. It is a growing process in which gaseous molecules collect in regularly oriented patterns on the outside surface of a solid material. The process is electrically and thermally induced. Epitaxy differs from diffusion in that the epitaxial interface is grown, whereas the diffused interface is the result of a molecular formation. An epitaxial grown layer is more easily process-controlled and possesses certain advantageous electrical properties, but is more expensive.

An epitaxial layer is usually grown over a P wafer as one of the initial steps in the formation of integrated P-N structures. This layer takes on the same crystal orientation as the single-crystal wafer. P isolation diffusions split the epilayer into sections or wells for the individual components. The epitaxial N layer is the collector of NPN transistors. The P-N junction formed by the epitaxial grown layer of N-type material upon a P-type substrate is also used extensively as part of the component isolation necessary for ICs.

**A Conclusion** The planar processes we have just described are complex, time-consuming, and proprietary in part. The explanation pro-

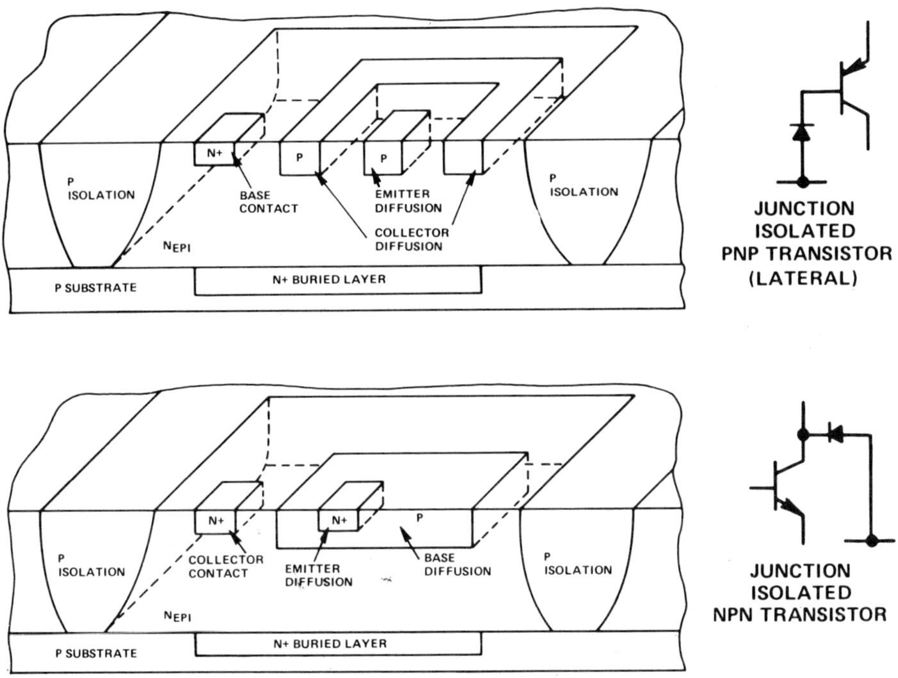


Figure 3-14 Key Diffusions in the IC Process (Courtesy, Analog Devices)

vided is cursory. The author would suggest that the reader choose a book such as *Semiconductor and Integrated Circuit Fabrication Techniques*, by Peter E. Gise and Richard Blanchard of Fairchild Semiconductor. This book is detailed and complete. It is published by Reston Publishing Co., Reston, Virginia.

Photomasking The process of photomasking is used to transfer the circuit image to the surface of the silicon wafer. It involves two subprocesses: the creation of the mask and the transfer process called photolithography.

**Creation of the Mask** Creation of the mask is begun with the desired circuit placed on a breadboard. The circuit is thoroughly tested to ensure that it produces the desired effects. The circuit is made of discrete components. Tests are made with changes in voltages, temperatures, and signals. The circuit then begins the process of transfer to a photomask. The photomask takes the following steps:

1. Parts are drawn in the geometry of integrated circuits.
2. Parts are arranged to fit the smallest possible space on the

device while allowing for interconnection and access to the outside world.

3. The engineering drawing is broken down into layers for the manufacturing process.

The layers just mentioned are photographically reduced and transferred to a master glass plate. A master is made for each layer. The masters are further made into submasters (working plates) used to perform the image transfer. The plates are covered with a photosensitive emulsion or other materials such as chrome, silicon, or iron oxide.

The fabrication of an IC requires from five to as many as twenty different masks, with six a typical number. A six-mask set is shown in figure 3-15. The masks are used one at a time, during each of the processing steps. Each one consists of a unique pattern established by a photographic emulsion on a flat glass plate. Areas of the photomask are either transparent or opaque to light. The masks are the same size as the wafer and must be highly accurate to define the proper regions for processing. This high degree of accuracy and resolution is achieved through a photographic process. As you can see in the figure, the buried layer is used with the others following in succession.

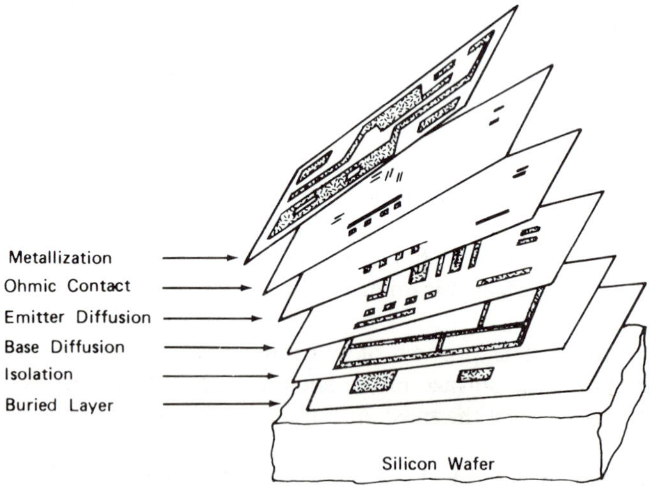


Figure 3-15 Photomask Set Used to Fabricate Integrated Circuits (Courtesy, Wojslaw, Reston)

**Photolithography** As was previously stated, the photolithographic process transfers the circuit image from the mask to the silicon wafer. As we have seen, the photosensitive material that allows this process to happen is called photoresist. Photoresist is a solvent base chemical mixture containing light-sensitive material. The purpose of the photoresist is to make possible the selective etching and removal of oxides or metals on the silicon wafer while protecting the remainder of the wafer and its parts.

The photolithographic process is accomplished in a series of steps. These are as follows:

1. Preparation of the wafer. The wafer is either manufactured by the IC manufacturer or purchased from a wafer company. In any event, the wafer arrives as pure material (either N or P type) with a known resistivity. It must be perfectly flat and devoid of scratches, cracks, or other deformities. The wafer is thoroughly cleaned with chemicals, then placed into an oven for oxidation. The silicon chemically reacts with the oxygen of water vapor to form an oxide layer on the silicon wafer. The thickness is easily controlled. In the figure 3-16, the oxide layer has been formed on a wafer of P-type material. This oxide layer is extremely thin, around 10,000 angstroms (which relates to $10^{-4}$ microns). The wafer is then coated with photoresist. The method used is usually a choice of the particular IC manufacturer. Some methods are spraying, dipping, or brushing. The most popular is the spinner. A photograph of a computerized photoresist coater (spinner) is shown in figure 3-17. Here the wafer is held by vacuum on a center shaft while the shaft spins. An exact amount of photoresist is deposited on the center of the shaft. The photoresist moves outward, leaving a coat of specified thickness on the wafer. Excess photoresist drops off the edge of the wafer. Thickness is a function of the shaft speed and the viscosity of the resistivity. The wafer is then soft-baked in an oven to remove the solvents and to harden the photoresist. Very precise optical equipment is used to align the photomask with the wafer. Alignment, as you can imagine, is also a critical step. The alignment of all the masks must be coordinated with previous mask alignments. After alignment, high-intensity ultraviolet light shines through the mask. This exposes the resist to the lamp rays in the areas not protected by the mask.
2. After exposure, the light and mask are removed and the wafer is placed in a developer solution. The areas unprotected by the polymerized photoresist are then dissolved. Edges where the photoresist has been removed are sharp. A rinse is used to ensure removal of any residue. The oxide layer is exposed in selected areas through the photoresist (see figure 3-18).
3. The wafer can now be thoroughly inspected and checked for alignment. It is then hard-baked. The photoresist is made more secure to the wafer and any solvent left in the photoresist is evaporated. The wafer is then immersed in an etching solution

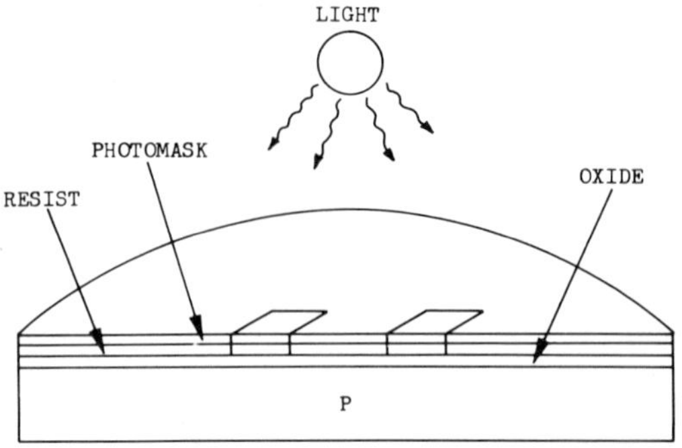


**Figure 3-16** The First Steps in Wafer Preparation (Courtesy, Analog Devices)

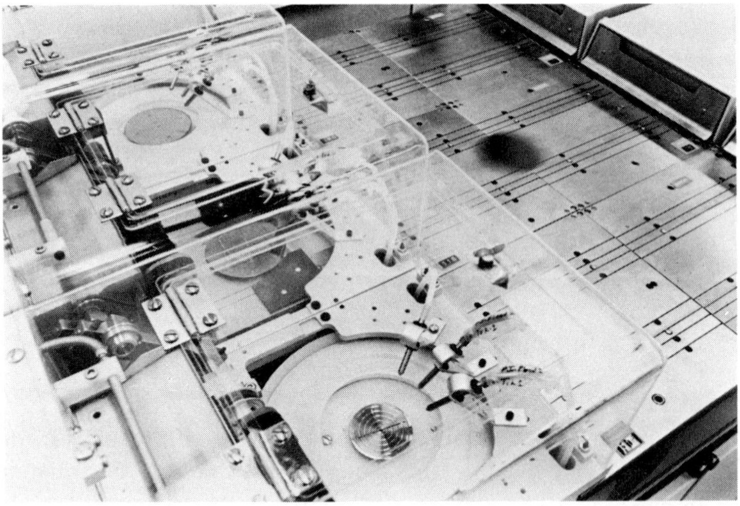

**Figure 3-17** Computerized Photoresist Coater (Courtesy, Analog Devices)

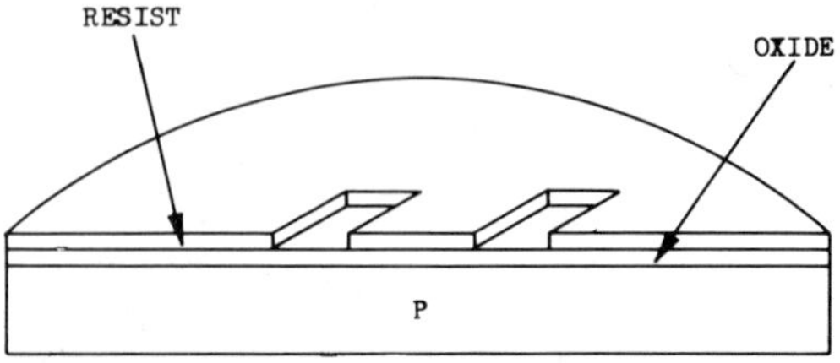


**Figure 3-18** Selective Removal of Photoresist and Exposure of the Oxide (Courtesy, Analog Devices)

such as hydrofluoric acid, which dissolves the oxide through exposed areas in the photoresist. The etching step is declared by some as the most critical of all the steps in the process (see figure 3-19).

4. After the etching process, the photoresist must be removed. It is dissolved in a solvent, or chemically removed in hot baths. The final product of this masking is shown in figure 3-20. The result of the process has made openings in the oxide layer where diffusion will take place.

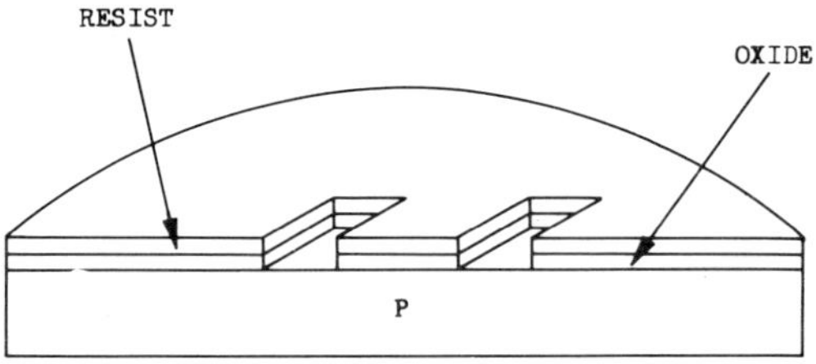


Figure 3-19 The Etched Oxide with Photoresist Still Intact (Courtesy, Analog Devices)

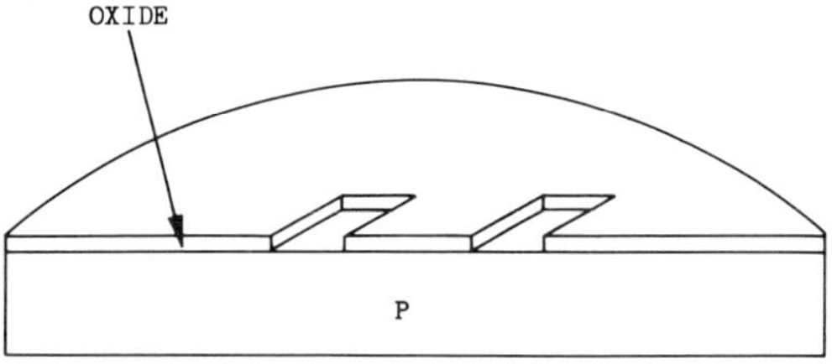


Figure 3-20 The Etched Oxide with Photoresist Removed (Courtesy, Analog Devices)

**The Fabrication of a Bipolar Device on a Wafer** As was shown in the discussion on photomasking, windows were made in the oxide for purposes of later diffusion or for epitaxial growth. Diffusion is the intermingling of dopants such as arsenic with a silicon substrate. The arsenic dopant is made to penetrate the silicon by means of temperature and concentration. Epitaxial growth is the growth of one crystalline material onto another that has the same physical structure.

**The Diffusion Process** The diffusion process can be likened to pouring milk into coffee. At entry, there is a high concentration of milk. With time, the milk mingles with the coffee and the coffee becomes lighter. After a long enough time, the milk is distributed throughout the coffee. The movement of the milk at the entry point to the rest of the coffee is called *diffusion*.

The process of diffusion on a silicon wafer is not unlike the coffee diffusion, but is dependent on the rate of the particles of dopant. The velocity of diffusion is dependent on temperature. There are two steps in diffusion. The first of these is *predeposition.* This method exposes an excess of the desired dopant at the surface of a silicon substrate such as P-type silicon. This is heated, and the material used as a dopant will enter the substrate until a maximum concentration of impurity is attained. The maximum concentration is called *solid solubility*, and is dependent on temperature alone. The second step in diffusion is *drive-in.* This method adds no additional dopant; it simply grows a protective layer of silicon dioxide over the fresh diffusion. The layer is carefully controlled for depth, thickness, and dopant profile.

**Buried-Area Diffusion** In the figure 3-21, an arsenic diffusion (N+) has been made through the windows of the oxide. The windows were created by means of the photoresist process. This heavily diffused area is the buried layer. The buried layer forms a low resistance region under most active devices such as the NPN transistor. It is often called the subcollector.

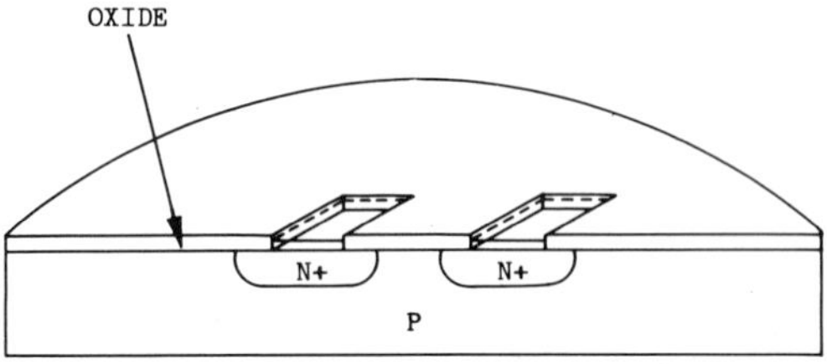


Figure 3-21 Buried Layer Diffusion (Courtesy, Analog Devices)

**Epitaxy** Epitaxy is the growing of one semiconductor onto another. Before epitaxial growth, the oxide mask still on from the diffusion process is removed with hydrofluoric acid. The wafer is then placed in a chamber called an epireactor, where the epitaxial area is grown over the P-type silicon and also over the diffused buried region. Control is carefully made with dopant semiconductors, temperature, and electricity (see figure 3-22).

**Second Photomasking and Isolated Area Diffusion** The wafer is returned to an oxidizing chamber and a layer of oxide is once more placed over the entire wafer. The oxide layer is about 1 micron thick. The wafer goes through the second photoresist operation (in this case, isolation masking) and windows are opened up for the second diffusion (P + boron). The isolation region creates separate areas in the epitaxial N layer for each of the IC components. After diffusion, there is a completely isolated N-type epilayer for the transistor we are fabricating (see figure 3-23).

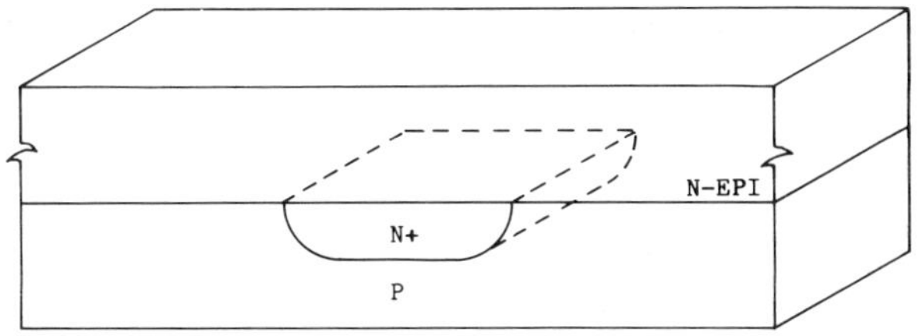


Figure 3-22 Epitaxial Growth Layer (Courtesy, Analog Devices)

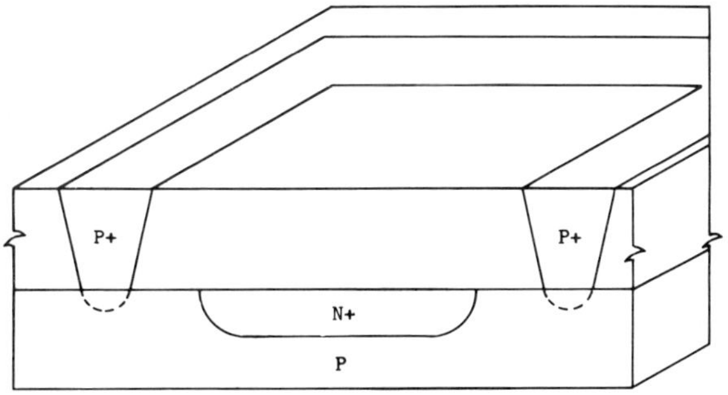


Figure 3-23 The Isolated EPI Area (Courtesy, Analog Devices)

**Third Photomasking and Base Diffusion** The wafer oxide mask is once again removed with hydrofluoric acid. It is then returned to the oxidizing chamber for another mask of 1-micron-thick oxide. The wafer goes through the third photoresist operation (in this case, the base diffusion) and windows are opened up for the third diffusion (P + boron).

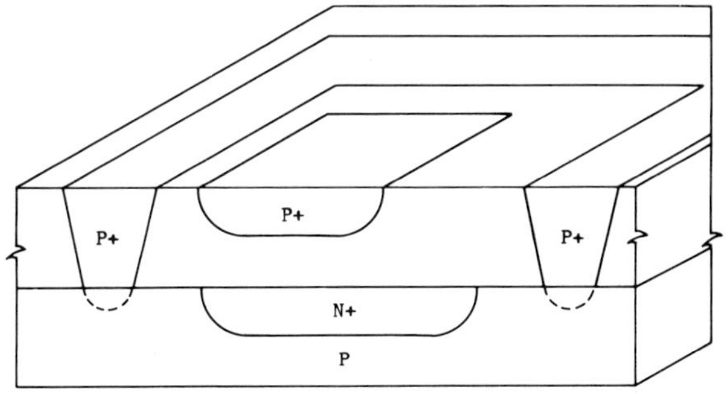


Figure 3-24 The Base Diffusion (Courtesy, Analog Devices)

The base diffusion is not nearly as deep as the isolation diffusion (3 microns). The base diffusion is made directly over the buried layer (see figure 3-24).

**Fourth Photomasking and Collector and Emitter Diffusion** For a fourth time, the wafer oxide mask is removed with hydrofluoric acid, another layer of oxide is applied, and the wafer goes through the photoresist process to create windows for the collector and the emitter (N + diffusion). In figure 3-25, the diffused area over the base is the emitter and the one to the side is the collector. The difference between the depths of the base and the emitter diffusions determines the base width which, in turn, determines the current gain (beta) of the transistor.

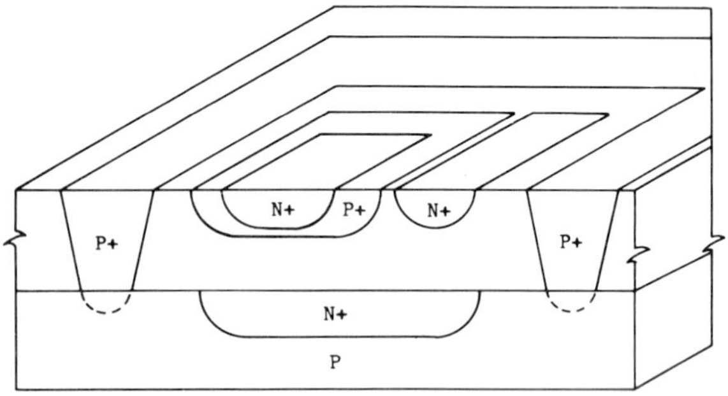


Figure 3-25 Collector and Emitter Diffusion (Courtesy, Analog Devices)

**Contact Masking** Again the oxide is removed and a new coat applied. The photoresist operation takes place to open windows directly above the emitter, base, and collector diffusions (see figure 3-26).

**Metallization** The old oxide is removed, a new coat is applied, and windows are opened by the photoresist process (see figure 3-27).

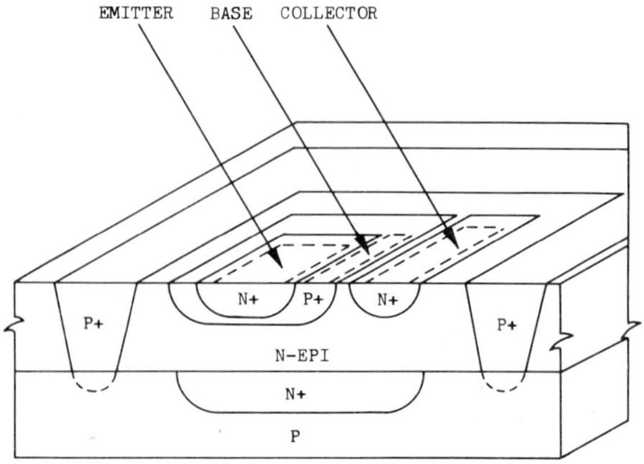


Figure 3-26 Contact Masking (Courtesy, Analog Devices)

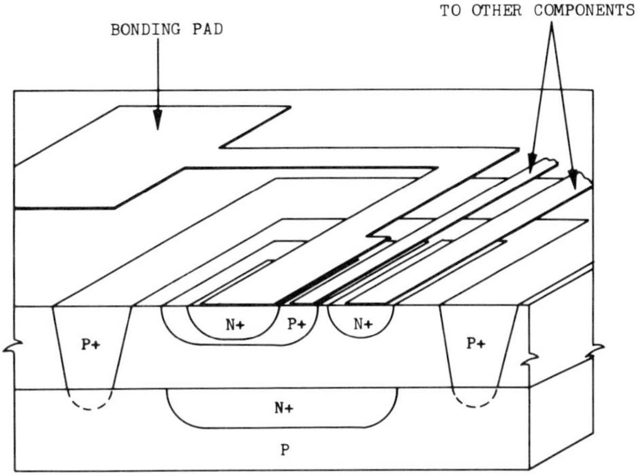


Figure 3-27 Metallization (Courtesy, Analog Devices)

Integrated-circuit components are interconnected with metallic, usually aluminum, traces during the metallization phase of the fabrication process. Also during this phase component and circuit bonding pads are formed. The component contacts must be ohmic or nonrectifying, and the circuit pads are provided for the bonding of small wires from the die to the package's leads. Metallization occurs through a process called *deposition.* During deposition, aluminum is evaporated in a high-vacuum system, and enough aluminum is permitted to deposit over the entire wafer surface. The photoresist process is again employed to protect the metal in the contact and interconnection areas, while the remainder of the metallization is exposed for removal by etching. The mask and process are inverted from those used in the diffusion steps.

There are five techniques used for metallization deposition: filament, electron beam, flash evaporation, induction evaporation, and sputtering.

*Filament evaporation* takes place from a filament or a boat by resistance heating. Current is gradually increased through the filament, which melts selected material. The liquid from the selected material flows on the filament and, with an increase in current flow, is evaporated. The vapor is then deposited on the wafer.

A high-intensity beam of electrons is focused on aluminum in a crucible to melt the aluminum. This is the *electron beam method.* It is a low-contamination process because nothing touches the material except electrons. Radiation damage may occur to the substrate because of the intense beam.

*Flash evaporation* utilizes a filament such as in filament evaporation. However, flash evaporation uses pellets, powder, or a spool of wire placed incident to the filament. This method is fast and free of contamination and radiation.

*Induction evaporation* uses a radio frequency source to couple power into metal in a crucible. R-f energy melts the metal, resulting in evaporation.

*Sputtering* is one of the most recent methods developed to deposit metal. Ions of inert gas are placed in a chamber at a vacuum. An electrical field is applied. The atoms are ionized. Ions strike a target in the chamber, dislodging atoms. The atoms are then deposited on a substrate in the chamber. Almost any material can be deposited, but at a low rate.

Final Wafer Processing After the photomasking, some final steps are needed to complete the integrated circuit. A cursory look at these processses follows.

**Probing** The last step in wafer processing is to test the die. At this stage, a machine called a *wafer prober* makes contact with the die through minute, pointed probes. These probes touch the aluminum pads of the IC and provide the electrical connection from a test instrument or system to the IC. Each die is electrically tested against predetermined specifications. Those dice that are determined to be faulty are inked and rejected.

**Dicing** (See figure 3-28) After the dice have been tested, they are ready to be separated by using a procedure called *dicing*. The most common method is scribing and breaking. Fine, diamond-cut lines are scribed vertically and horizontally across the wafer but between each circuit or die. The wafer is then placed on a rubber pad, and pressure from a roller is exerted on the wafer, breaking it up into individual chips.

**Assembly** Assembly is the most expensive part of the fabrication process. In wafer processing, large numbers of circuits are processed simultaneously, but during assembly the dice are handled individually, and connections to each die are made separately. Assembly operations

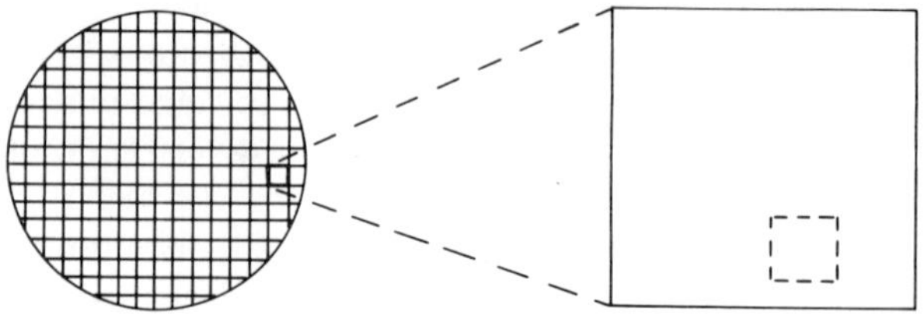


Figure 3-28 Dicing (Courtesy, Analog Devices)

are required to protect the die, to facilitate handling, and to provide a package and contacts for equipment application.

**Die Bonding** Die bonding is the soldering, brazing, or glassing of the die to a package frame. This attachment provides a mechanical and, occasionally, an electrical contact from die to package. This bond also acts as a thermal path allowing internally generated heat to flow to the surroundings.

**Wire Bonding** The electrical connections from the chip to the package's terminals are made by one of several wire bonding techniques. One-mil (0.001-inch) wire, either gold or aluminum, is connected to an IC aluminum pad and then extended to the package terminal, where a similar connection is made. Ultrasonic, pressure, or temperature techniques are used to form the wire to pad and terminal welds. Great emphasis is being placed on eliminating this costly and labor-oriented method and replacing it with an automated procedure. Presently, most IC manufacturers have shifted to an automated bonding technique by which connections are made from die to the package by the use of metal tabs.

**Packaging** The individual IC die is too small and delicate to handle. To facilitate handling and protect it from damage, the die is welded to a frame, its leads are connected, and the unit is sealed, passivated, and encapsulated. Glass, ceramic, and plastic are common materials that encase the die. There are many types of IC packages, but the most popular are the axial lead (T05), flat-pack, and dual-inline (DIP) packages.

**Testing** The last step in the manufacturing process is the electrical testing of the IC. DC, AC, and functional measurements are made to ensure that the IC will perform to established standards. Large-volume products are automatically transported by mechanical autohandlers. A computer-controlled IC test system, in conjunction with the autohandler, systematically and quickly screens each IC to verify its performance. Bad ICs are rejected, and those accepted are marked and stored for distribution.

Summary of Integrated-Circuit Processing (See figure 3-29) It must be noted that all integrated-circuit fabrication processes using the planar process are similar. Differences lie in the functions of the devices, such as transistors (bipolar) and MOS devices (unipolar). Of course, other differences lie in the number of devices on the chips. In the figure, the major process steps are illustrated. Minor steps have been omitted for purposes of clarity. The reader may also be reminded that each step in the fabrication process is important to the manufacturer for economic and quality purposes. Each step is proprietary and carefully guarded. It may take years to perfect a technique that will save time or material. These special techniques are essential to compete in the IC field.

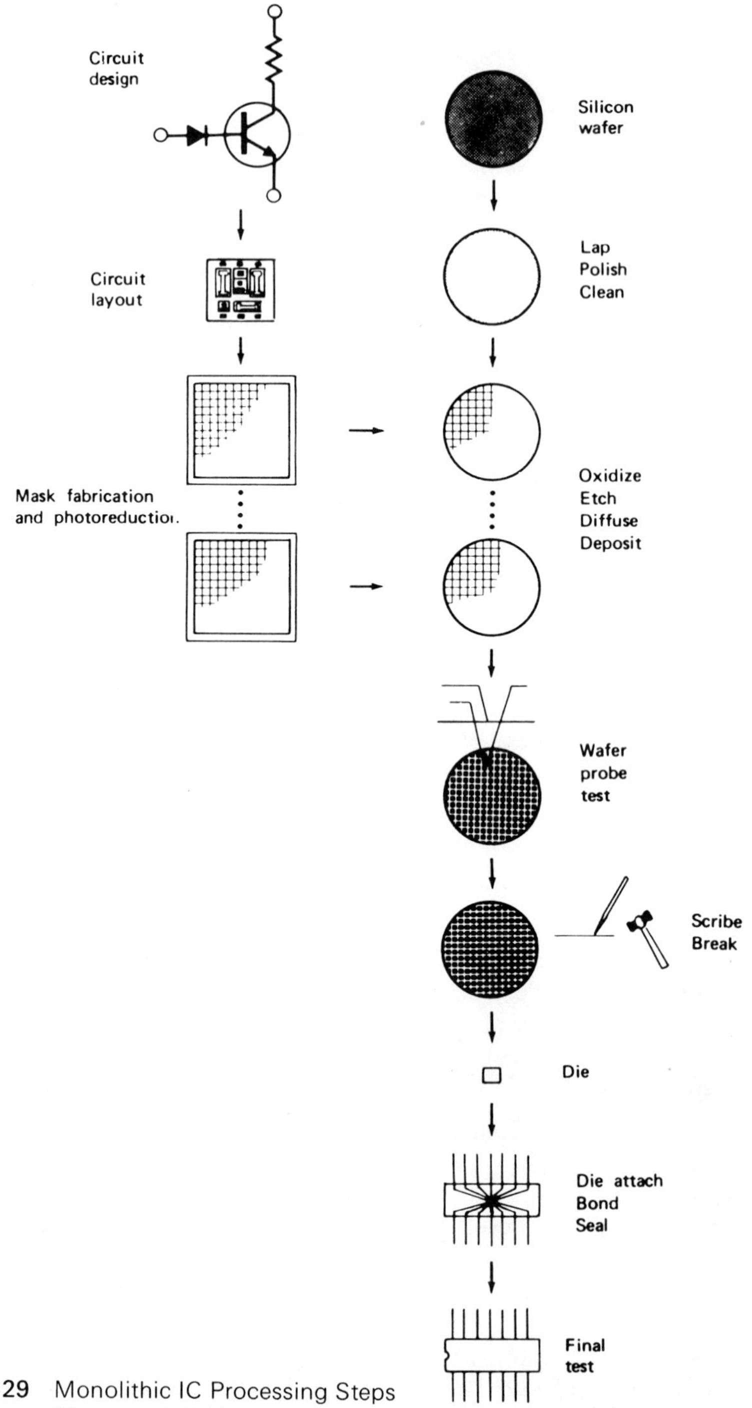


Figure 3-29 Monolithic IC Processing Steps (Courtesy, Wojslaw, Reston)

# 4
# Operational Amplifier Fundamentals

## AN INTRODUCTION

An operational amplifier IC is a solid-state integrated circuit that uses external feedback to control its functions. The operational amplifier may be used for a great number of circuits, which include amplifiers, integrators, filters, differentiators, voltage followers, oscillators, regulators, and mathematical circuits. Its greatest characteristic is simplicity in operation.

The operational amplifier can replace entire transistor amplifier circuits in many applications such as audio, radio, analog, and signal-conditioning systems. Its basic operation depends on the use of negative feedback, as described later in this book.

Each operational amplifier has unique qualities that can be determined from the manufacturer's specifications. Some of these qualities are wide frequency range, internal frequency compensation, low power consumption, internal "short" protection, and temperature stability. Operational amplifiers are, in fact, the integrated-circuit evolution to the differential amplifier. They are especially noted for their high input impedance ($10^{12}$ Ω), high voltage gain ($10^6$), low output impedance (10 Ω), and zero output signal for zero input signal.

**The Typical Operational Amplifier** (See figures 4-1 and 4-2) The $\mu$A741 is typical of the operational amplifier. It is particularly useful for schools, in that it is a general-purpose type of device and available at low prices. The internal circuit of the $\mu$A741 operational amplifier is illustrated in figure 4-1. Figure 4-2 shows the pad layouts for the device.

The reader will note that the figure 4-1 is an equivalent circuit for the $\mu$A741 operational amplifier. Transistors Q1 and Q2 are the noninverting and inverting inputs, respectively. They are differential inputs.

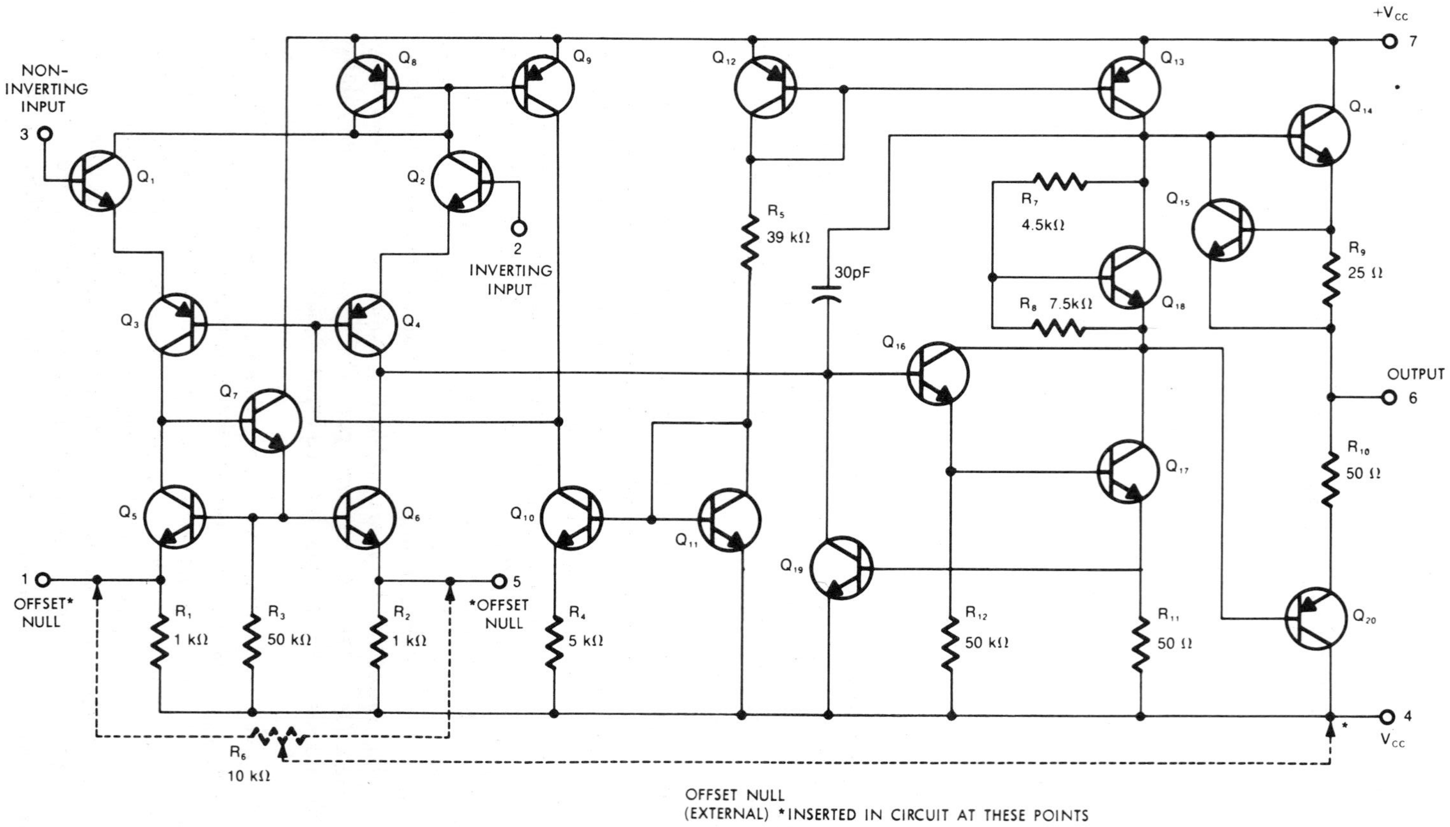


**Figure 4-1** Equivalent Circuit for $\mu$A741 Operational Amplifier (Courtesy, Fairchild Semiconductor)

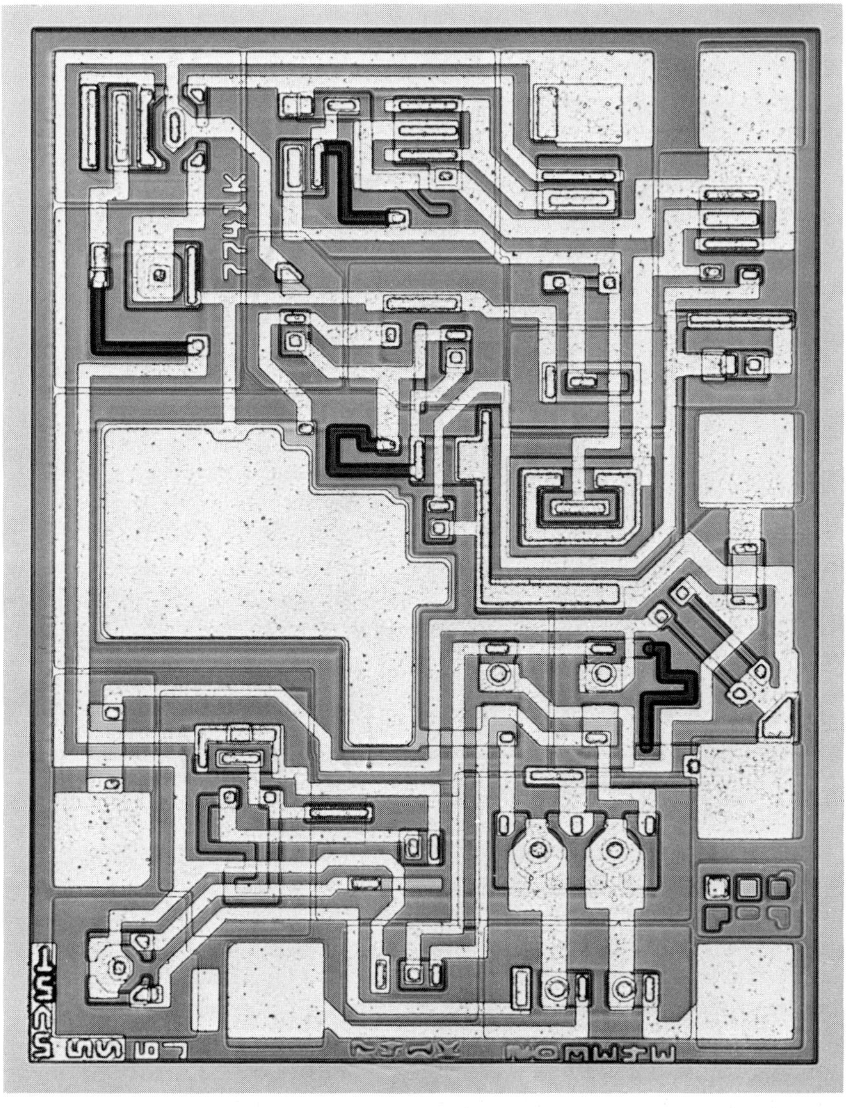

**Figure 4-2** Pad Layout for $\mu$A741 Operational Amplifier (Courtesy, Fairchild Semiconductor)

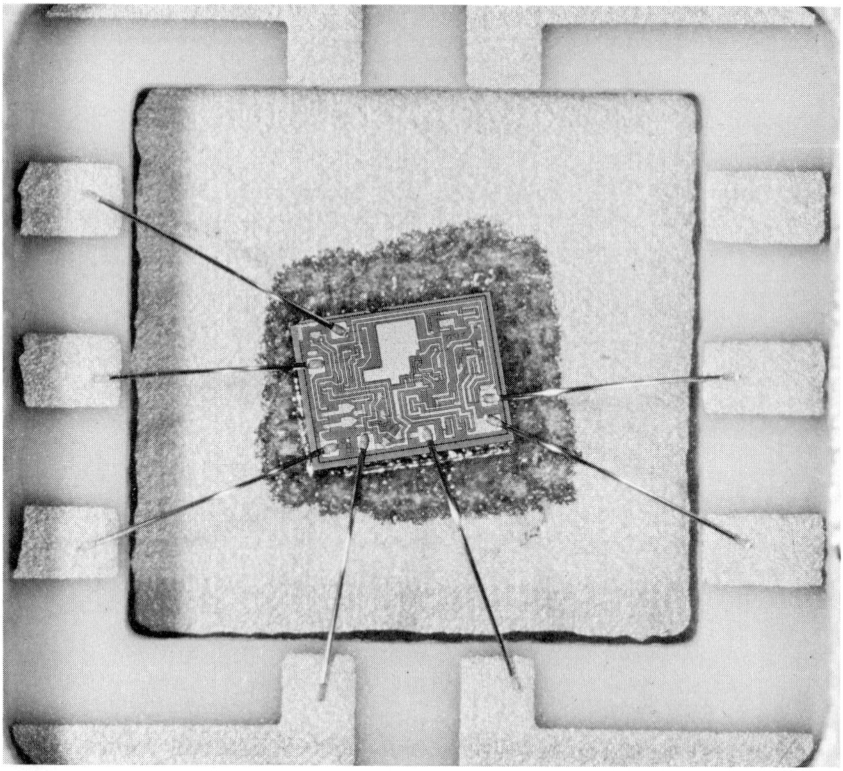

Figure 4-2 continued

Transistor Q18 is a high-gain Darlington driver. The transistors Q14 and Q20 form a class AB complementary symmetry output stage. The rest of the transistors and components provide bias and amplification.

Null offset adjustment is made by placing a 10 K potentiometer between pins 1 and 5 with the wiper tied to $-V_{CC}$. Short-circuit protection is supplied by current-limiting transistor Q15 in the output stage.

Since this book is written for operational amplifier design applications, we shall leave study of transistor theory to the reader. In figure 4-2 the reader will also note the actual $\mu$A741 pad layout and connection diagram. It is interesting to compare the circuit in figure 4-1 with the pad layout in figure 4-2.

As a final comparison of related items, a drawing of the $\mu$A741 is shown in its 8-pin form T-case in figure 4-3. Shown along with the operational amplifier are a typical mounting pad and socket.

Terminal Designations (See figure 4-4) The figure illustrates terminal designations for typical operational amplifier packages. Each

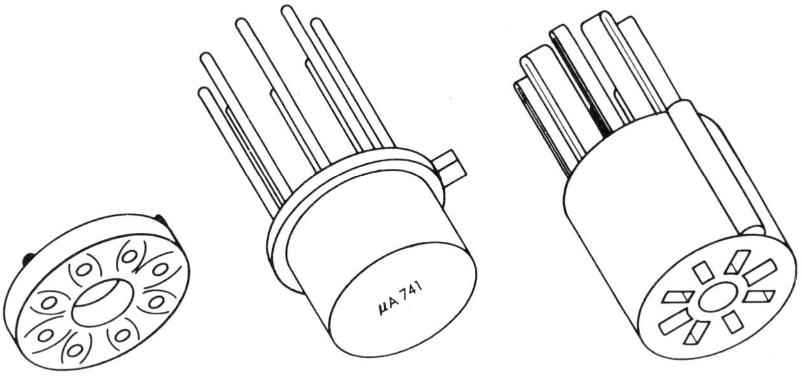


**Figure 4-3** Typical μA741 Operational Amplifier in an 8-Pin Form T-Case

package contains inverting and noninverting inputs. Each package also contains negative and positive power-supply inputs. Some packages contain offset null positions. Some packages contain two complete operational amplifiers.

The terminals and their designations are not standardized throughout the industry. Therefore, before design or use, the manufacturer's specifications must be thoroughly studied to ensure that the correct pins are used for the function desired.

**Basic Principles of Operation** (See figure 4-5) The principle of feedback in the operational amplifier IC is the function that makes it so versatile. This principle is quite simple. A feedback connection is made between the output and inverted input leads, with resistors used as voltage dividers. The voltage divider provides the exact amount of negative feedback for any level of amplification (gain) required or desired.

Closed-loop gain is dependent on the size of the feedback resistance. Open-loop gain (no feedback resistance) can be ignored in most designs, but is used in voltage-comparison and level-detection circuits.

As was stated in the discussion on feedback, the gain of the operational amplifier is controlled externally. The operational amplifier is said to have infinite gain within the restraints of power supplied at pin 7 and 4. It is also said to have infinite input impedance and zero output impedance. When there is no signal applied to either the inverting (pin 2) or noninverting (pin 3) inputs, the output (pin 6) is zero. When a signal is applied to one input, the opposite input strives to approach that signal level so that there is always a differential of zero between the two inputs.

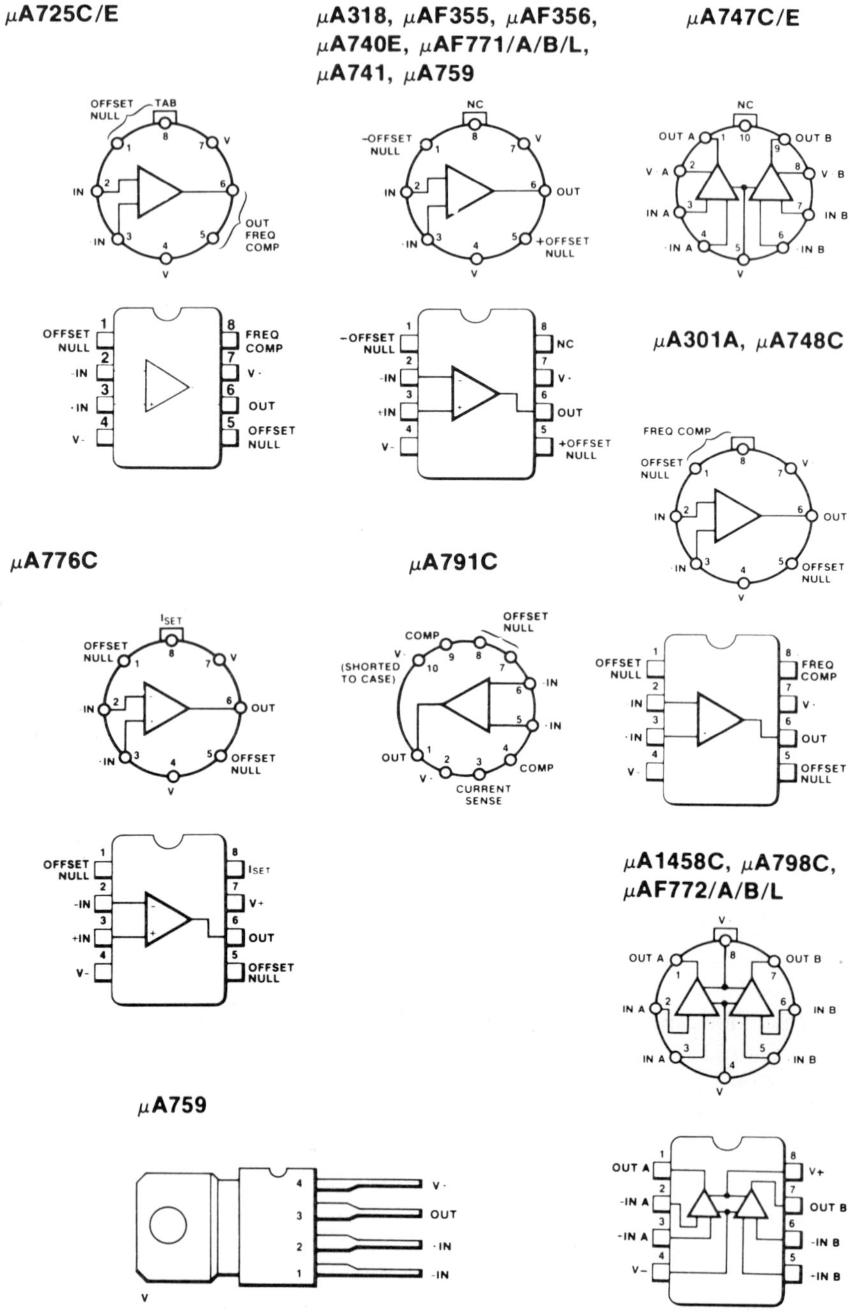


**Figure 4-4** Terminal Designations (Courtesy, Fairchild Semiconductor)

## Linear

### μA709C

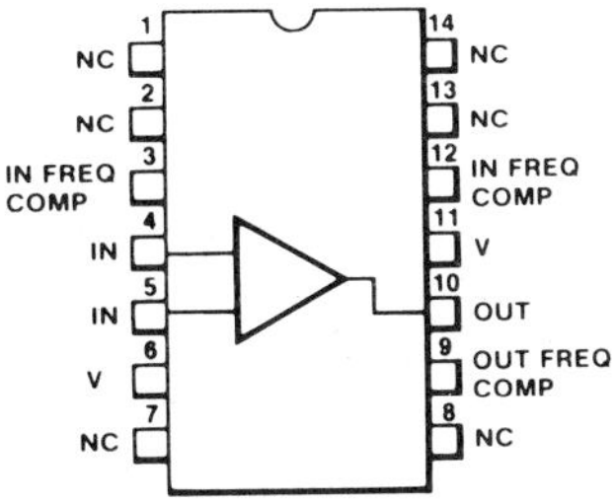


### μA715C

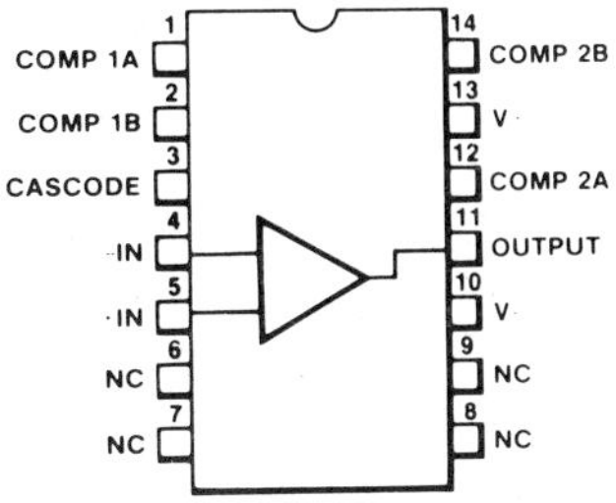


### μA747C/E

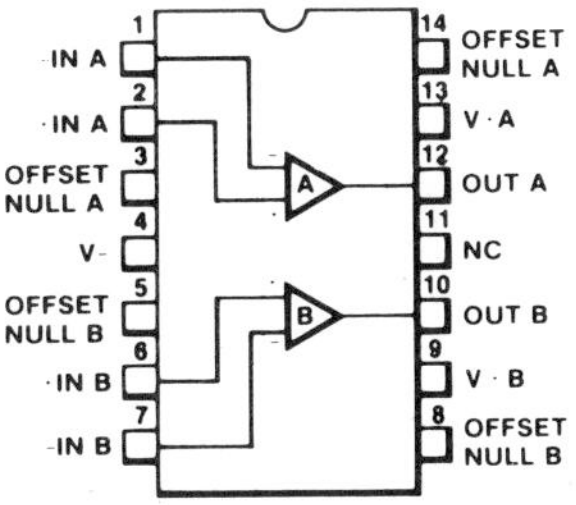


### μA324, μA348, μA3403
### μA774/A/B/L

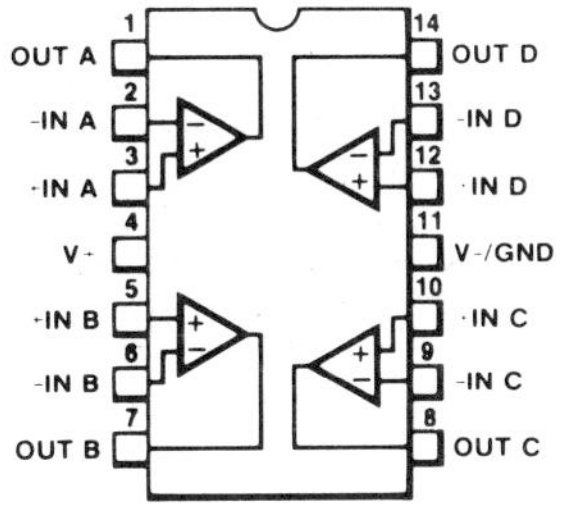


Figure 4-4 continued

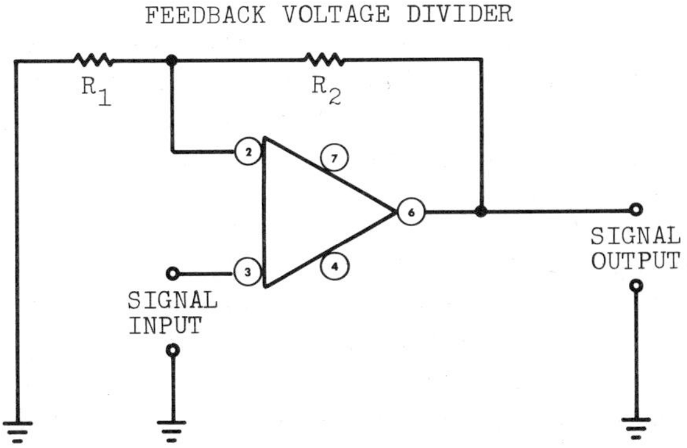


Figure 4-5 Operational Amplifier: Principles of Operation

Since feedback is from output to inverting input, the output will always be that amount necessary to drive the two inputs to near null (or an extremely small value). Current is so small at the inputs that it is assumed to be nonexistent. Output polarity is in phase with the noninverting input and out of phase with the inverting input.

More detailed operation is provided later in the book.

**The Differences Between Operational Amplifiers** Each operational amplifier is constructed to perform at its best under certain conditions. Families of operational amplifiers are designed to optimize a set of parameters crucial to particular applications. Usually the titles in the operational amplifier catalogs provide some indication as to the set of parameters that best utilizes the specific "skills" of an op amp. Some of these are listed below. This list was taken from Data-Acquisition Products Catalog, Analog Devices, Inc.

1. General purpose (low cost)
2. FET-input, low bias current
3. Electrometers (lowest bias current)
4. High accuracy, low drift differential
5. Chopper amplifiers (lowest drift)
6. Fast wideband
7. High output
8. Isolated op amps

Each of the sets of parameters is used under circumstances pertinent to the particular application. For instances, items 1, 2, and 4 are most relevant for transducer applications, while item 7, high output, may be used in power applications.

**Operational Amplifier Versatile Elements** Probably the reason that the operational amplifier has such wide application is its versatility. It seems to be the workhorse of linear electronics. It is equally useful in digital applications. For the purpose of providing the reader with food for thought, a list of some of the more useful elements is provided. There are literally dozens more.

1. High gain and low input excursion
2. Negative input terminal must follow the positive terminal
3. Output voltage must be whatever is necessary to achieve step 2
4. The device can generate inverse functions—that is, the gain is equal to the inverse of attenuation of the feedback circuit
5. Conversion of current to voltage is possible
6. Conversion of voltage to current is possible

7. The operational amplifier can establish virtual ground: voltage null and current balance
8. Inputs can be summed independently
9. The operational amplifier can function as a sign inverter
10. The operational amplifier can activate passive circuitry, linear or nonlinear, static or dynamic, single-element or network
11. The operational amplifier can handle differential signals

## OPERATIONAL CONCEPTS

**The Ideal Amplifier (See figure 4-6)** The search for an ideal amplifier is, of course, an exercise in futility. The characteristics of the operational amplifier are good enough, however, to allow us to *treat* it as ideal. Following are some amplifier properties that make this so. (The reader must realize that these ratings are seldom, if ever, reached.)

1. Gain—infinite
2. Input impedance—infinite
3. Output impedance—zero
4. Bandwidth—infinite
5. Voltage out—zero (when voltages into each input are equal)
6. Current entering the amplifier at either terminal—extremely small

For analysis purposes, the voltage gain of an operational amplifier is extremely high (the $\mu$A741C reaches $A_v = 150{,}000$). Typical input impedances vary between 1 and 50 megohms, with some modern devices

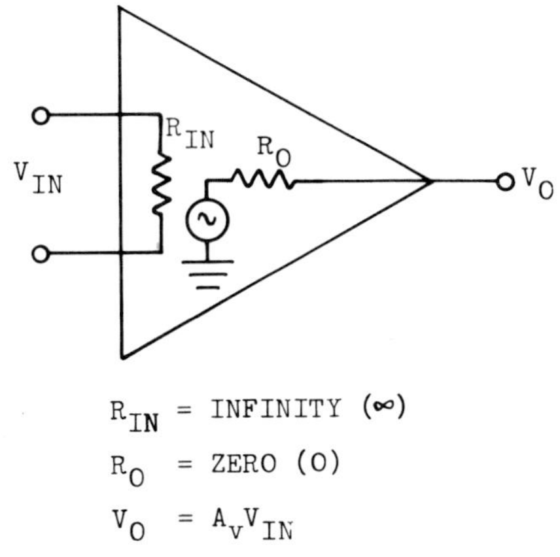


Figure 4-6 The Ideal Amplifier

reaching the mega-megohm level. Output impedances are generally a bit less than 75 ohms. The bandwidth may range from 1 to 5 megahertz.

The reader will probably understand infinite gain, infinite input impedance, zero output impedance, and infinite bandwidth from discrete amplifier studies.

The ideal amplifier has a zero voltage out only when both inputs are equal. If there is even the slightest voltage difference between the two input terminals, the output will become large in comparison. This is, of course, due to the great voltage gain that the operational amplifier can attain.

Since the input impedance approaches infinity, the input current is small and leaning toward infinitesimal.

**Open-Loop Operation** (See figure 4-7) Consider the open-loop response of the amplifier in the figure. Note that there is no external feedback. Pin 2 is the inverting input and the point at which the signal is input. Pin 3 is the noninverting input. Pin 3 is grounded. Pin 6 is the signal output. Since the gain is extremely high (approaching infinity), the difference in the two inputs must be as near zero as possible. In the ideal amplifier this is the case.

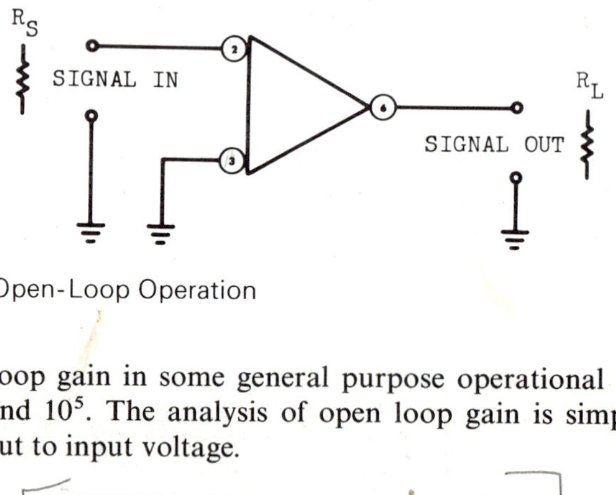


Figure 4-7 Open-Loop Operation

Open-loop gain in some general purpose operational amplifiers is usually around $10^5$. The analysis of open loop gain is simple. It is the ratio of output to input voltage.

$$\frac{\text{SIGNAL OUT}}{\text{SIGNAL IN}} = \text{Open-Loop Gain}$$

With no signal there is no input. Therefore at pins 2 and 3 the voltage is zero. At pin 6 the voltage is also zero. If pins 2 or 3 take any signal value, voltage at pin 6 increases to saturation.

A small amount of signal such as 3 millivolts will cause the operational amplifier to traverse to the active region. This is the same effect as in a switching transistor circuit; the output is either high (1) or low (0).

Open-loop operation is used in voltage comparators. Other applications are control-loop-gain scheduling, tracking monitors, and level detection. However, open-loop operation is not nearly as versatile as closed-loop operation.

Feedback Principles (See figure 4-8) The reader may recall that the open-loop gain of an operational amplifier is extremely large—say, 200,000. Consider placing a signal into the amplifier with a gain of this size. If the maximum output voltage $V_{OUT}$ of an amplifier is $\pm 10$ volts with a gain of 200,000, the largest voltage that could be applied would be $\pm 50\ \mu V$. The effects of feedback can lower this gain and control it. Most amplifier situations do not require large gains but do require gain stability. By placing external feedback resistance on the amplifier, we can control the gain, somewhat reduce the output impedance, and in some configurations increase the input impedance.

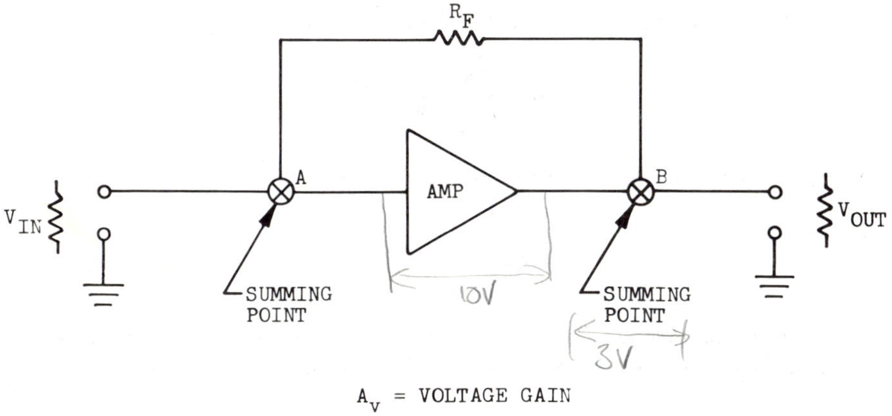


Figure 4-8 Feedback Principle

In figure 4-8, an amplifier is illustrated with a feedback resistor $R_F$. $V_{IN}$ is the input signal. $V_{OUT}$ is output voltage. Joints $A$ and $B$ are summing points. The amplifier has $A_v$ (voltage gain).

The summing point $B$ is in series with the amplifier's input. Any voltage at $B$ will also be felt at summing-point $A$. Summing-point $A$ will also feel any voltage from the input $V_{IN}$. The summing-point $A$ then feels the difference voltage between the two sources:

$$V_A = V_{IN} - V_B$$

The output voltage $V_{OUT}$ is the function of the difference voltage $V_A$ and the voltage gain:

$$V_{OUT} = A_v V_A$$

or

$$V_{OUT} = A_v(V_{IN} - V_B)$$

**A Closed-Loop Amplifier (See figure 4-9)** In closed-loop operation, the voltage gain of an operational amplifier is determined by the feedback resistance $R_F$. The resistance $R_M$ is a voltage divider resistance (metering resistance in instrumentation) that is selected to match the impedance of the source. Voltage difference between input terminals 2 and 3 approaches zero. The input voltage $V_{IN}$, in this configuration, is placed at pin 2.

As we discussed in the section entitled "Feedback Principles," the voltage at $A$ was the sum of the voltage in $V_{IN}$ and the voltage at summing-point $B$. Such is the case for the current. Current through the feedback resistance $R_F$ and the input resistance $R_M$ is summed at point $A$ ($I_A$).

The current $I_{RF}$ is determined by Ohm's law:

$$I_{RF} = \frac{V_{OUT} - V_A}{R_F}$$

The current $I_{RM}$ is also determined by Ohm's law:

$$I_{RM} = \frac{V_{IN} - V_A}{R_M}$$

At the summing point of the amplifier, current is considered approximately zero because the only path to ground is through the high

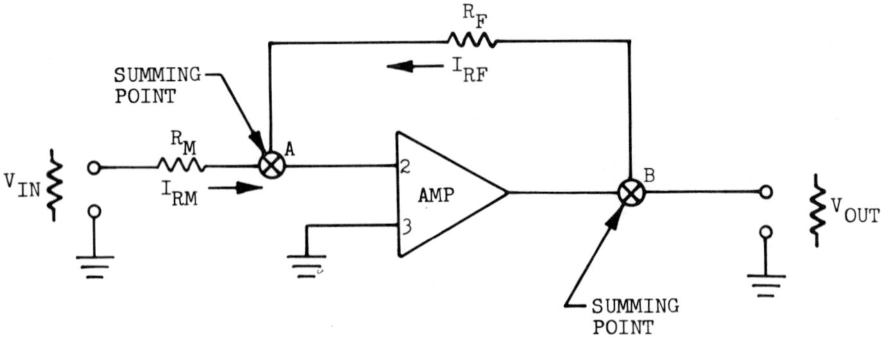


$$A_V = -\frac{R_F}{R_M}$$

Figure 4-9 Closed-Loop Operation

impedance of the amplifier. Therefore the current from the source and the feedback are the same:

$$I_{RM} = I_{RF} \cong I_A$$

The association of voltage and voltage gain was previously discussed as a direct proportion:

$$V_{\text{OUT}} = A_v V_A$$

Voltage at the input summing-point $A$ is $V_A$. $V_A$ is considered virtual ground, since it usually is extremely small (as is current). The resistors $R_F$ and $R_M$ provide a voltage divider from output to input. If a positive $V_A$ is applied to this inverting amplifier, the voltage at the summing-point $A$ will rise above zero. This will drive the output voltage negative because the signal was applied to the inverting input. The output will increase negative until the voltage at the summing-point $A$ returns to zero. $R_F$ and $R_M$ are therefore a voltage divider between the input $V_A$ and $V_{\text{OUT}}$.

It seems at this point necessary to list the equalities that exist:

$$V_{\text{OUT}} = A_v V_A$$

$$I_{RF} = \frac{V_{\text{OUT}} - V_A}{R_F}$$

$$I_{RM} = \frac{V_{\text{IN}} - V_A}{R_M}$$

$$I_{RF} = I_{RM} \cong I_A$$

$$\frac{V_{\text{IN}}}{R_M} = -\frac{V_{\text{OUT}}}{R_F}$$

$$\frac{V_{\text{OUT}}}{V_{\text{IN}}} = -\frac{R_F}{R_M}$$

$$\therefore \quad A_v = -\frac{R_F}{R_M}$$

## KEY SPECIFICATIONS

All specifications are, of course, critical to an application. Some are more critical than others. In any event, the first choice of key specifications will narrow the field of requirements when you are making your op amp choice.

**Input Offset Voltage ($V_{OS}$)** Input offset voltage is a small voltage (microvolts to millivolts) that appears in series with the input leads of an op amp. It is that voltage required at the input to obtain a zero-output voltage. The cause of the voltage appearing at the inputs is a mismatch of the base emitter voltages of the first-stage input transistors or, in the case of the FET input amplifiers, the gate source mismatch. It is undesirable voltage and must be considered in your amplifier choice. You may simply choose an op amp that provides an offset voltage lower than required, or you may null the voltage.

Nulling may be achieved by placing a voltage at an input that cancels out the offset. The second method of nulling is to place an imbalance offset voltage of opposing polarity within the amplifier. This will result in actual offset of the amplifier. Terminals are provided on the op amp for this adjustment. Note that these terminals are to be used only for trimming the offset. They should never be used to compensate for other errors in the amplifier.

**Offset Voltage Drift** "Input offset voltage" is not a fixed, predictable term. It varies with time, temperature, and even an unstable power supply. Power supplies are usually well regulated and this term is not too much of a concern. Temperature, however, may vary the offset voltage a few microvolts. Some amplifiers drift in time. Drift with time is a rarity and is not cumulative. Drift is that input offset voltage that still exists when the output has been adjusted to zero. The designer may choose to select an op amp with a very low drift.

**Input Bias Current ($I_B$)** An ideal amplifier has zero current. Such is not the case in real-life situations, however. Bias current must flow at the input terminals of the op amp. In some events, the current is the actual bias current in the first stage of the op amp. In other cases it is leakage current. The problem that arises from the current is that it flows through the external resistors, causing voltage drops and rates of voltage change in capacitors. If you are lucky, the bias current on each differential input will be equal. This allows affected resistors on the inputs to be the same, thereby nulling the effects of this current.

Input bias current is defined as the average of the input dc biasing current:

$$I_B = \frac{(I_B^+) + (I_B^-)}{2}$$

Its value varies with op amp type, but can reach 1.5 $\mu$A. Input bias current changes with changes in temperature.

**Input Offset Current** Input offset current is the difference between the input bias currents at the inverting and the noninverting inputs. That is,

$$I_{OS} = (I_B^+) - (I_B^-)$$

The difference $I_{OS}$ is generally smaller than either $(I_B^+)$ or $(I_B^-)$. This parameter simply suggests how well-matched the input currents really are.

The cause of the offset is the imperfect match between the betas of the input transistors. Typical values for input offset current range from 1 nA to 100 nA.

**Input Offset and Bias-Current Drift** Stable bias currents will in turn allow for stable offset. Likewise, a drifting input bias current may cause voltage drift. Currents tend to be extremely low in op amps—therefore current compensation is not required. Drift is usually caused by an excessive temperature. If compensation is made, it is at one temperature.

**Common-Mode Voltage (CMV)** Common-mode voltage (also called common-mode signal) is a voltage that is the same amplitude and phase. This signal is applied to both the inverting and the noninverting input terminals. An ideal op amp reacts only to difference voltage between the two input terminals. However, since ideal is seldom achieved, the designer must reckon with common-mode voltage. Owing to variations of sensitivity of the inverting and noninverting inputs or variations of offset voltage, this voltage may be amplified and appear at the output.

**Note:**

$$V_{IN}(\text{inverting}) = V_{IN}(\text{noninverting}) = \text{CMV (or } V_{CM})$$

**Common-Mode Gain (CMG) or $A_{CM}$)** This is the gain of the operational amplifier with a common-mode voltage (CMV) applied to the op amp input terminals. A common-mode gain is much less than one (say, 0.01).

**Common-Mode Error (CME)** If this same voltage (CMV) is amplified and felt at the output, then referred to the input by feedback (dividing by a voltage gain), the error is said to be the common-mode error (CME). The CME is effectively in series with the inputs.

**Common-Mode Rejection Ratio (CMRR)** Common-mode rejection affects an op amp's ability to deal with common-mode signals. CMRR is a ratio of common-mode voltage (CMV) gain to open-loop gain. The ratio is called common-mode rejection ratio (CMRR).

CMRR is a ratio and is expressed as such (example 20,000):

$$\text{CMRR} = \frac{\text{Open-Loop Gain}}{\text{Common-Mode Gain}} = \frac{A_{OL}}{A_{CM}}$$

CMRR is also expressed logarithmically:

$$\text{CMRR (in decibels)} = 20 \log_{10} \text{CMRR}$$

For example, a typical CMRR would be 86 dB.

Data in specifications relating to CMRR are usually average figures. CMRR decreases with increasing frequencies.

**Input Impedance ($Z_{IN}$)** Input impedance of an operational amplifier is said to approach infinity. In actuality, each device does have a typical input impedance such as 1 megohm and above. Some of the more sophisticated devices may reach an input impedance of as much as 100 megohms. This value may be determined from the manufacturer's specifications.

The input impedance ($Z_{IN}$) of an op amp is the resistance at either the inverting or the noninverting inputs with the opposite input at ground.

**Output Impedance ($Z_{OUT}$)** The output impedance of an operational amplifier is said to approach zero. In actuality, each device is different, but for the sake of design this impedance can be assumed to be zero. In specifications, output impedances are said to be 25 to 50 ohms. However, this value can be ignored for most applications.

Because of low output impedance, the output will function as a voltage source and will provide required current for a wide range of loads up to the current limit level (25 mA, typically). With this capability and high input impedance, the operational amplifier becomes an ideal impedance-matching device.

**Slew Rate** Slew rate is the rate of change in the output of an operational amplifier for a square wave input. The slew rate (amplification time) of an operational amplifier is very fast. A typical example is 0.5 volt per microsecond. Slew rate in operational amplifiers may vary from low (say, 0.1 volt per microsecond) to high (say, 100 volts per microsecond).

Fast slew rates become important for high frequency and fast switching applications. For most applications, however, the slowest slew rate is more than satisfactory. Manufacturer's specifications provide the slew rate for each device.

**Frequency Response** Figure 4-10 is called a Bode plot. This plot is used to compare gain and frequency. The term "decade" is often used in connection with frequency. A decade is a frequency change of ten times. For instance, a change from 10 to 100 hertz would be a decade. Likewise, a change from 100 to 1,000 hertz would be a decade.

Since the operational amplifier has an open-loop gain of 100 dB for frequencies less than 10 hertz (also dc), a typical closed-loop gain of 20 dB will extend the high frequency ($f_H$) point to 100,000 hertz (the gain drops 80 dB over a range of 4 decades). The relationship between voltage gain ($A_v$) and dB is: dB $= 20 \log_{10} A_v$. The shaded area in figure 4-10 represents an amplifier with a gain of 20 dB. The slope is 20 dB per decade or 6 dB per octave. The point at which the gain begins to fall off is called falloff ($f_o$). The point at which the gain drops to zero is called unity or $f_t$.

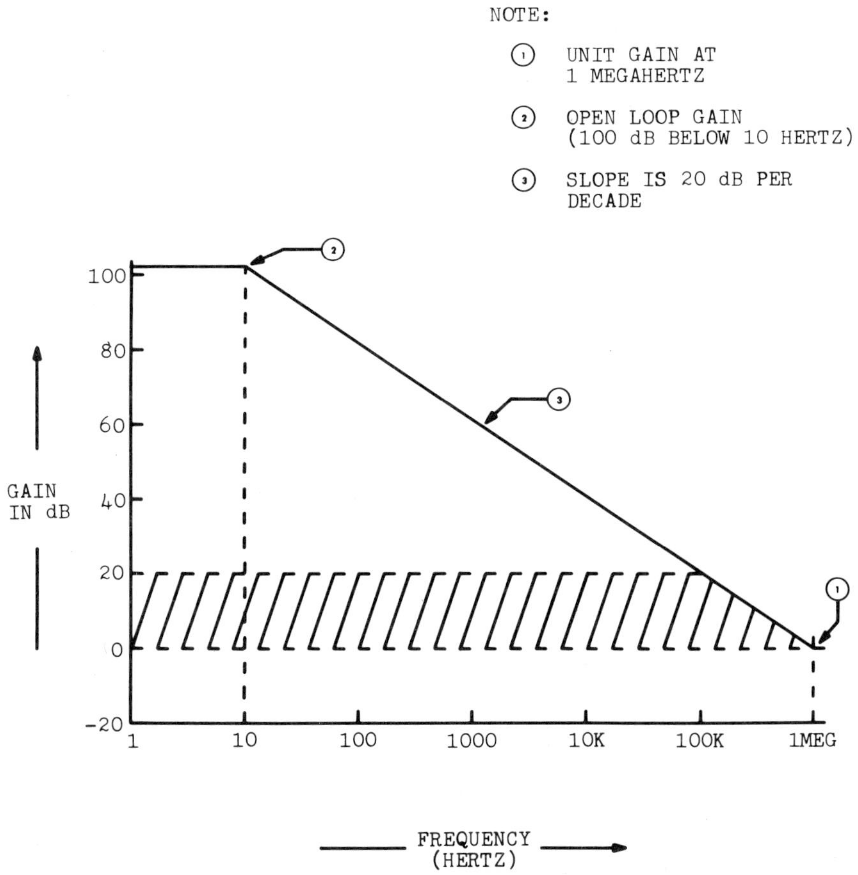


Figure 4-10 Gain (In dB) as a Function of Frequency

**Noise in the Amplifier** Operational amplifier noise is that level of output which is inherent in an amplifier with no input ($V_{IN}$) signals. This assumes that the operational amplifier input has a resistive path to ground and the output ($V_{OUT}$) signals have been nulled by adjustment of the offset null potentiometer.

Since noise is predominantly a high-frequency term, it is possible to improve the signal-to-noise ratio with the use of filtering and by reducing amplifier gain as a function of frequency, to ensure amplifier stability. The latter method is termed *compensation* and is especially important where extremely high gains are achieved, such as with the operational amplifier.

From experience, it has been determined that reduction in gain of 20 dB per decade is optimum compensation for most operational amplifier applications. Compensation in the operational amplifier is accomplished internally or externally. For special circuits, an uncompensated amplifier can be selected, and it is up to the designer to choose the desired compensation.

Filtering, on the other hand, can be accomplished by placing a capacitor on the input or output to shunt undesired frequencies to ground. It should be noted that noise is more pronounced in high-impedance circuits, and lead length should be a consideration in design applications.

## GAIN

Probably the most important one parameter/specification involved with an amplifier is gain. In fact, that is what the entire exercise is all about. An output is derived from some input. The amplifier provides an output that is "X" amount larger than the amount placed into the input.

Voltage gain with operational amplifiers approaches infinity for the ideal amplifier. For non-ideal amplifiers, the voltage gain is adjustable with external resistances. In either event, the following voltage gain ratio exists:

$$A_v = \frac{V_{OUT}}{V_{IN}}$$

where $A_v$ = voltage gain

and $V$ = voltage

This may seem rather basic; however, it is essential to most amplifier situations. Two other gain ratios, $A_i$ (current gain) and $A_p$ (power gain), should be stated as fundamental to the amplifier cause:

$$A_i = \frac{I_{\text{OUT}}}{I_{\text{IN}}}$$

where $A_i$ = current gain

and $I$ = current

$$A_p = \frac{P_{\text{OUT}}}{P_{\text{IN}}}$$

where $A_p$ = power gain

and $P$ = power

All three of these equations are simply ratios of one value to another and are, of course, dimensionless.

All other parameters and specifications evolve around modifications to the gain ratios. The modifications are either desirable or undesirable.

**Open-Loop Voltage Gain** Open-loop gain is the voltage gain of the op amp without a feedback but with a load. These values are around $10^5$ (not infinity) and are dependent on frequency. The open-loop dc power gain $A_p$ is around 100 dB (again, not infinite).

Open-loop gain may vary with temperature, voltage supply, and configuration load. Indeed, open-loop gain may vary somewhat with each device within a specified part number. Open-loop gain will always vary with frequency.

**Feedback Attenuation ($\beta$)** The term "feedback attenuation" ($\beta$) is a ratio of signal feedback to signal output:

$$\beta = \frac{R_M}{R_M + R_F}$$

where $R_M$ = input resistance (metering)

$R_F$ = feedback resistance

The two resistances represent the total signal attenuation. The ratio value can approach 1/1 (unity) where there is zero attenuation. In this event, there would be 100 percent feedback.

Feedback attenuation is often called feedback factor.

**Loop Gain** Loop gain is the total gain of the loop or the product of the loop. That is the voltage gain ($A_v$) times the feedback attenuation ($\beta$). In formula, this is:

$$\text{LG} = A_v \beta$$

where

$\text{LG}$ = loop gain

$A_v$ = voltage gain open loop

$\beta$ = feedback attenuation (factor)

**Signal Gain** Signal gain is closed-loop gain and is a simple ratio for both inverting and noninverting input signals. For inverting stages the signal gain is

$$A_v = -\frac{R_F}{R_M}$$

where

$A_v$ = voltage gain

$R_F$ = feedback resistance

$R_M$ = metering resistance

For noninverting amplifiers, the gain is

$$A_v = \frac{R_M + R_F}{R_M}$$

**Noise Gain** Noise gain is the reciprocal of the attenuation factor ($\beta$) for the inverting configuration:

$$\text{noise gain} = \frac{1}{\beta} = \frac{R_M + R_F}{R_M}$$

It is important to remember that noise gain for the noninverting configuration is the same.

# 5
# Preliminary Design Considerations

The difficulty with design of any electronic circuit is that the results of theoretical design are not always satisfactory. That is, the real-life circuit may not perform as you have anticipated. This is no reason to bury your head in the sand. All designers run into the same problems. Some are more intense. However, most can be controlled. This chapter is devoted to some of the design problems and their solutions.

## CHOOSING AN OP AMP

The choice of an operational amplifier is obviously important. The task is quite simple. Each manufacturer publishes complete specifications for each of its operational amplifiers. See table 5-1 for specifications on the $\mu$A741 operational amplifier, which is used extensively as the design example in this book.

It is up to the designer to select a device that meets his circuit needs. Basically, the designer should determine his load, gain requirements, impedances, power availability, job type, and packaging. He should then go to the manufacturer's specifications to find an operational amplifier that meets these needs.

The $\mu$A741 is frequency-compensated. Other operational amplifiers (such as the $\mu$A709) are available without compensation, but these have provisions for external compensation. It is usually desirable to use external compensation for high-frequency applications.

**Table 5-1** Design Specifications for $\mu$A741 Operational Amplifier

**ELECTRICAL CHARACTERISTICS** ($V_S = \pm 15$ V, $T_A = 25°C$ unless otherwise specified)

| PARAMETERS (see definitions) | CONDITIONS | MIN. | TYP. | MAX. | UNITS |
|---|---|---|---|---|---|
| Input Offset Voltage | $R_S \leq 10\ k\Omega$ | | 1.0 | 5.0 | mV |
| Input Offset Current | | | 20 | 200 | nA |
| Input Bias Current | | | 80 | 500 | nA |
| Input Resistance | | 0.3 | 2.0 | | MΩ |
| Input Capacitance | | | 1.4 | | pF |
| Offset Voltage Adjustment Range | | | ±15 | | mV |
| Large-Signal Voltage Gain | $R_L \geq 2\ k\Omega,\ V_{out} = \pm 10$ V | 50,000 | 200,000 | | |
| Output Resistance | | | 75 | | Ω |
| Output Short-Circuit Current | | | 25 | | mA |
| Supply Current | | | 1.7 | 2.8 | mA |
| Power Consumption | | | 50 | 85 | mW |
| Transient Response (unity gain) | $V_{in} = 20$ mV, $R_L = 2\ k\Omega$, $C_L \leq 100$ pF | | | | |
| Risetime | | | 0.3 | | μs |
| Overshoot | | | 5.0 | | % |
| Slew Rate | $R_L \geq 2\ k\Omega$ | | 0.5 | | V/μs |
| The following specifications apply for $-55°C \leq T_A \leq +125°C$: | | | | | |

Table 5-1 *continued*

| | | | | | |
|---|---|---|---|---|---|
| Input Offset Voltage | $R_S \leq 10$ kΩ | | 1.0 | 6.0 | mV |
| Input Offset Current | $T_A = +125°C$ | | 7.0 | 200 | nA |
| | $T_A = -55°C$ | | 85 | 500 | nA |
| Input Bias Current | $T_A = +125°C$ | | 0.03 | 0.5 | μA |
| | $T_A = -55°C$ | | 0.3 | 1.5 | μA |
| Input Voltage Range | | ±12 | ±13 | | V |
| Common Mode Rejection Ratio | $R_S \leq 10$ kΩ | 70 | 90 | | dB |
| Supply Voltage Rejection Ratio | $R_S \leq 10$ kΩ | | 30 | 150 | μV/V |
| Large-Signal Voltage Gain | $R_L \geq 2$ kΩ, $V_{out} = \pm 10$ V | 25,000 | | | |
| Output Voltage Swing | $R_L \geq 10$ kΩ | ±12 | ±14 | | V |
| | $R_L \geq 2$ kΩ | ±10 | ±13 | | V |
| Supply Current | $T_A = +125°C$ | | 1.5 | 2.5 | mA |
| | $T_A = -55°C$ | | 2.0 | 3.3 | mA |
| Power Consumption | $T_A = +125°C$ | | 45 | 75 | mW |
| | $T_A = -55°C$ | | 60 | 100 | mW |

(Courtesy, Fairchild Semiconductor)

## DEALING WITH MINUS SIGNS IN DESIGN PROCEDURES

In using the design procedure equations for operational amplifier circuits, the beginning designer may be disturbed by an occasional minus sign for the value of some component such as a resistor. At first sight this appears to make no sense at all, yet when one multiplies or divides a minus value by a plus value, the algebra clearly requires a minus product or quotient.

In most ordinary calculations, zero is taken as an absolute reference for all quantities; less than zero is negative and more than zero is positive. In electronic circuitry, however, this beautiful clarity is sometimes lacking. A voltage of +2 is less positive than +5, for instance, so the first voltage is considered negative where +5 is taken as the reference voltage.

Because the operational amplifier is composed of many discrete transistor amplifiers, it has the same characteristics they do. Some of these cause phase shifting, both lagging and leading. For example, a common-emitter transistor will cause an input phase shift of 180 degrees, or total reversal, because of the time consideration. Compared to ordinary experience, the time seems extremely short, but at electronic frequencies it may be long, as though the events of a year could be compressed into a few microseconds.

Also, most of the operational amplifiers utilize ±vdc. Power supplies must be used for transistor or FET biasing purposes. The use of ±vdc power supplies causes current to flow in negative and positive circuits leading in many directions toward some reference other than zero.

For these reasons, the minus signs have been placed in the design procedures of this book to denote that output signals are out of phase with input signals, and they serve no other purpose. In making calculations, the minus signs should be disregarded to avoid confusion.

## OFFSET NULLING

### Manufacturer's Offset Voltage Null Adjustment (See figure 5-1)

Since the operational amplifier is of the differential type, input offset voltage must be controlled so as to minimize offset. Offset voltage is nulled by application of a voltage of opposite polarity to the offset.

An offset null-adjustment potentiometer may be used to compensate for offset voltage. The null-offset potentiometer also compensates for irregularities in the operational amplifier manufacturing process which may cause an offset. Consequently, the null potentiometer is recommended for critical applications.

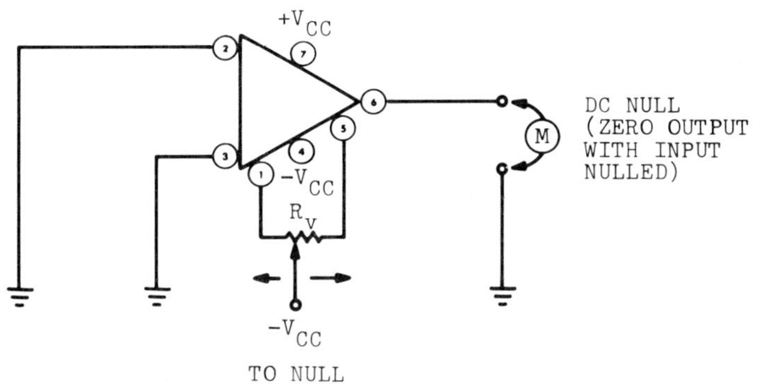


NOTE: BOTH INPUTS GROUNDED AS SHOWN ABOVE INPLIES THAT $V_{IN}$ IS ZERO AND THAT EACH INPUT HAS A DC RESISTANCE PATH TO GROUND. THE ACTUAL RESISTANCE DEPENDS ON THE APPLICATION.

Figure 5-1 Offset Null Adjustment

Offset adjustment is explained in the next paragraph. Design verification in this book will ask for offset null adjustments that may or may not be necessary for your design. Offset null-adjust potentiometers are not placed on design schematics, because they would serve to detract from the design.

Perform the following procedures:

1. Adjust a 10 K potentiometer to its center position.
2. Connect the potentiometer outside leads between pins 1 and 5 of the operational amplifier. Make sure that the power is applied to the design application.
3. Connect the wiper of the potentiometer to the negative supply voltage.
4. Ensure that input signals are zero or null and that pins 2 and 3 have a dc return to ground.
5. Measure the output with a dc meter and obtain zero null by adjusting the potentiometer.

This is just one offset null procedure. It is recommended by vendors for the design of the $\mu$A741 operational amplifier used in this book. Check other vendors' specifications for other methods of offset null adjustment. It must be noted that the preferable method of setting offset null is that which is suggested by the manufacturer of the device.

**Universal External Offset Nulling** Besides the adjustment recommended by the device manufacturer, there are some other methods that work just as well. Use the manufacturer's suggested method if possible. If you can't, for some reason or other, then one of the following procedures may serve as an alternative.

**Inverting Amplifier Offset Nulling** (See figure 5-2) The figure illustrates an inverting amplifier. Note that the signal is placed into the amplifier on pin 2. The offset is isolated from the signal input by being placed on the noninverting input pin 3.

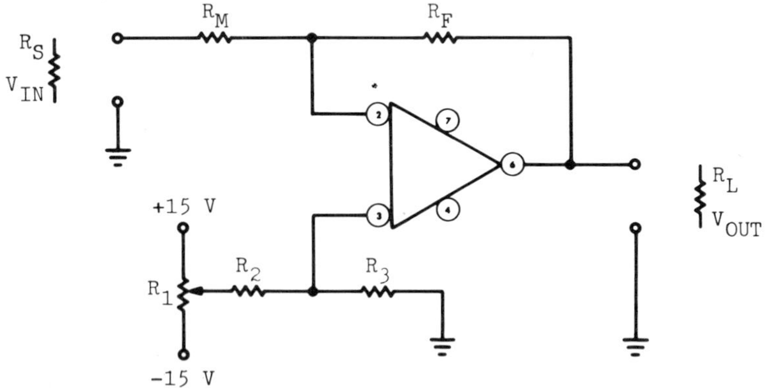


Figure 5-2 Inverting Amplifier Offset Adjustment

Perform the following procedures:

1. Choose a potentiometer of 25 K or 50 K (1 percent tolerance).
2. Choose a voltage divider $R_2$ and $R_3$ for 1,000 to 1 ratio (example, $R_2$ 100 K and $R_3$ 100 Ω) (1 percent tolerance).
   **Note:** As shown, this circuit will provide a $\pm 15$ millivolt offset voltage adjustment range.
3. Connect the voltage divider and apply power to the divider and op amp.
4. Make sure that the input signal is at zero.
5. Measure the output with a dc meter and obtain zero null by adjusting the potentiometer.

It must be noted that the potentiometer used may add noise to the amplifier. The best tolerance potentiometer and voltage divider resistors provide a better chance to avoid noise.

**Noninverting Amplifier Offset Nulling** (See figure 5-3) The figure illustrates a noninverting amplifier. The compensation network is placed

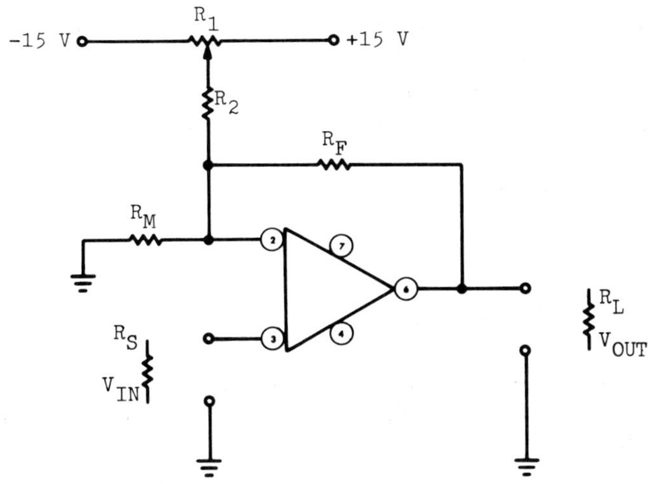


Figure 5-3 Noninverting Amplifier Offset Adjustment

at the summing point to the inverting input. As in the inverting amplifier offset nulling, the adjustment is isolated from the input signal.

Perform the following procedures:

1. Choose a potentiometer of 25 K or 50 K (1 percent tolerance).
2. Choose a resistor $R_2$ to be 1,000 times $R_M$ (1,000 to 1 ratio) (for example if $R_M = 100\ \Omega$, then $R_2$ should be 100 K) (1 percent tolerance).

   **Note:** As shown, this circuit will provide a $\pm 15$ millivolt offset voltage adjustment range.
3. Connect the voltage divider and apply power to the divider and op amp.
4. Make sure that the input signal is at zero.
5. Measure the output with a dc meter and obtain zero null by adjusting the potentiometer.

It must be noted that the potentiometer used may add noise to the amplifier. The best tolerance potentiometer and voltage divider resistors provide a better chance to avoid noise.

**Other Design Offset Nulling** The voltage follower and the differential amplifier are also candidates for offset nulling. Offset nulling will be covered under those headings.

**Trimming of Instrument Grade Op Amps** Special care should be taken in trimming instrument grade op amps. In the first place, this quality of op amp usually has low offset and low drift. Therefore, procedure for setting offset may indeed place more problems into the design than the reason for nulling did in the first place. The advice here is first to choose the op amp and be satisfied that it has low offset and drift. Second, if you must trim, use the procedure recommended by the manufacturer of the op amp. Third, use extremely accurate trim pots and/or resistances.

## FREQUENCY COMPENSATION

One of the problems that most general-purpose op amps (and others) have is the loss of gain at higher bandwidths. You may recall from our previous discussions on bandwidth versus gain curves that the op amp gain begins to roll off at very low frequencies in open-loop operation. (In most general-purpose op amps this is somewhere between 10 and 100 hertz.) Gain continues to fall off to unity at frequencies of 1 megahertz or so.

It stands to reason, then, that the design of a closed-loop amplifier should include some facility to compensate for loss of gain at higher frequencies. Along with that problem of loss of gain, we must also compensate for phase shift to achieve closed-loop stability. The method generally used is to prevent rolloff to unity gain before 180° phase shift is reached. If this does not occur, the op amp may go into oscillation. The use of a simple RC circuit reduces the phase shift to less than 90°. In this manner a margin is provided between 90° and the 180° to make room for parts tolerance and loading problems.

**Lag-Phase Compensation** Most modern op amps are internally compensated for a 6 dB/octave rolloff. Other op amps must be externally compensated. Terminals on the op amp are designated *phase*, *rolloff*, or *frequency compensation*. At these terminals an external RC network is connected to provide frequency compensation. Op amp manufacturers usually specify the RC combination to achieve lag compensation for a particular closed-loop gain. It is particularly advisable to follow their suggestions.

**RC Input Compensation** (See figure 5-4) One method of phase compensation is to selectively (at random) choose an RC network for which the capacitor has a lower reactance than the resistance at the chosen frequency. Place the two components in series across the inputs to the op amp. This method may add noise to the amplifier, but selectively will remove the oscillation.

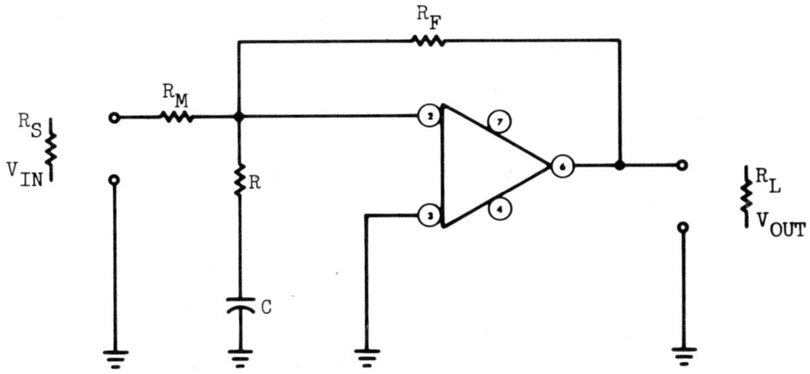


Figure 5-4 RC Input Compensation

Lead Compensation (See figure 5-5) A second method used to equalize phase shift is to place a capacitor in parallel with the feedback resistor. The choice is broad. Start with 100 pF and go upwards toward 1 $\mu$F or so. This method will lose gain at high frequencies.

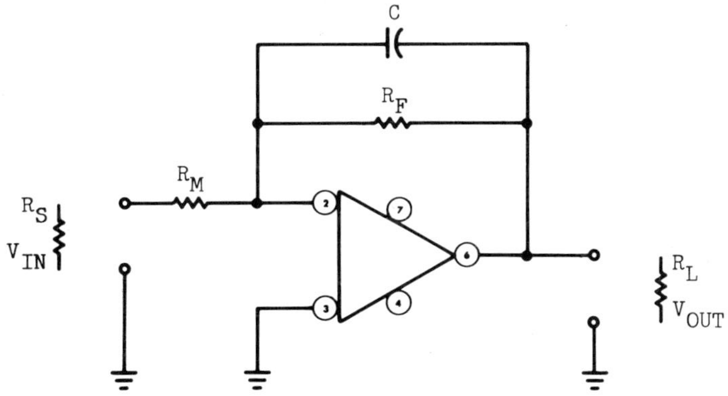


Figure 5-5 Lead Compensation

## INPUT BREAKDOWN

If the input voltage rating of an op amp is exceeded, the op amp will of course have internal failure. To prevent overvoltage failure, a pair of signal diodes are placed in parallel reversed from one another across the input leads. Then a resistor (say, 7.5 K to 15 K) is placed in series with each input. The diodes clamp the differential input to a safe level. The resistors limit the current into the inputs.

Input breakdown is not an extreme problem because most modern-day devices are protected with internal differential clamping devices.

## LATCHUP (See figure 5-6)

Some op amp configurations have a tendency to latchup. Latchup can occur in positive feedback configurations. It occurs when the output swing is very near the input signal. The op amp goes into saturation. If saturation occurs, the inverting stage no longer inverts and negative feedback is now positive feedback. The configuration is now latched up. Latchup can be broken by removing power.

To compensate for this phenomenon, a pair of diodes are placed across the input to the amplifier. Latchup occurs only at extreme inputs. Therefore, if extremes can be eliminated, latchup can also be eliminated.

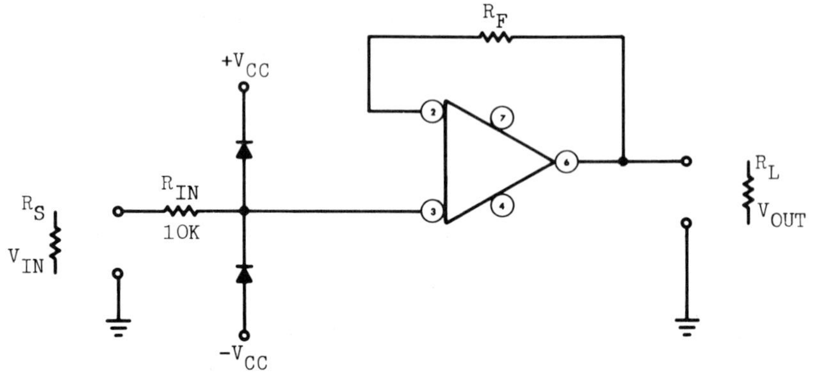


Figure 5-6 Latchup Compensation

## COMMON SYMBOLS AND ACRONYMS FOR OPERATIONAL AMPLIFIER DESIGN APPLICATIONS

Table 5-2 lists some of the common symbols and acronyms found in design applications for operational amplifiers. These are typical. Each designer or manufacturer may, of course, use his own symbols or just letters and numbers (for example, $R_1$, $R_2$, etc.).

**Table 5-2.** Common Symbols and Acronyms for Operational Amplifier Design Applications

| | | | |
|---|---|---|---|
| $A$ | Gain (Sometimes $G$) | $I_{RM}$ | Current Through Metering Resistance |
| $A_i$ | Current Gain | $R_F$ | Feedback Resistance |
| $A_{ol}$ | Open-Loop Gain | $R_G$ | Gain Setting Resistance |
| $A_v$ | Voltage Gain | $R_L$ | Load Resistance |
| $A_{vc}$ | Close-Loop Voltage Gain | $R_M$ | Metering Resistance |
| CME | Common-Mode Error | $R_N$ | Noninverting Input Resistance |
| CMRR | Common-Mode Rejection Ratio | $R_S$ | Source Resistance |
| | | $R_V$ | Voltage Divider Resistance |
| CMV | Common-Mode Voltage | $V_A$ | Voltage Applied At Summing Point (also $V_{IN}$) |
| $E_O$ | Output Voltage (also $E_{OUT}$) | | |
| $I_A$ | Current Applied at Summing Point | $V_{CC}$ | Supply Voltage (also $V_S$) |
| | | $V_{IN}$ | Input Voltage |
| $I_{IN}$ | Current from Source | $V_{OS}$ | Offset Voltage |
| $I_L$ | Load Current (also $I_{OUT}$) | $V_O$ | Output Voltage (also $V_{OUT}$) |
| $I_{OS}$ | Offset Current | $Z_{IN}$ | Input Impedance |
| $I_{RF}$ | Current Through Feedback Resistance | $Z_L$ | Load Element |
| | | $Z_{OUT}$ | Output Impedance |

# 6
# Design Procedures and Examples

Design, of course, is an exacting procedure. It involves specific mathematical exercises that are necessary for the building blocks of instrumentation mechanization. This chapter is dedicated to some basic design procedures and examples. The author has attempted to keep the design as simple as possible. The reader should have very little difficulty in coping with the simple applications presented.

Before we get into design, it may be intelligent to remind ourselves of several of the basic laws that govern the operation of the op amp. We may repeat these laws from time to time in this chapter.

1. No current flows into the amplifier ($I$ applied, $I_A = 0$).
   **Note:** Remember that the input summing point is virtual ground.
2. The difference voltage between inputs, inverting and noninverting, is zero (0).
3. With a feedback attached (closed-loop), the negative input signal will be driven to the positive reference input.

## DESIGN AND TEST CHECKLIST

This list presents only a few of the many things you can do to ensure that your design is successful. A satisfactory design may be thrown out because it has not been tested properly. Add to this list whenever you run into a problem that may repeat itself in other designs.

1. Make sure that power is connected to the proper supply points (for example, + on pin 7, − on pin 4).
2. Make sure that grounds are made properly. Doubling grounds often helps.

3. In the event of saturated output, check open feedback loop or open ground return on noninverting input.
4. If the operational amplifier you are working with has no internal short protection, take extra care to prevent shorts.
5. Isolate your operational amplifier with a pad or a socket to help prevent shorts.
6. Check the actual circuit against the schematic to ensure that parts are the right ones before you apply power.
7. Recheck calculations prior to applying power.
8. Make sure your test equipment is grounded prior to use on your circuit.
9. Attempt to use the same source for voltage and ground returns.
10. Attempt to prevent overloading input or output.
11. Make sure inputs are placed into whichever (inverting or noninverting) input that is decided upon.
12. Use a power supply that is regulated and filtered if at all possible.

## VOLTAGE FOLLOWER

The voltage follower is the simplest of the configurations of the op amp. Voltage followers using operational amplifiers are of two basic designs: dc and ac. Their purposes, in both types, is to transfer signals from large to small impedances with near-unity voltage gain from input to output. They are essentially impedance-matching circuits that control voltage output but allow current to the load.

**DC Voltage Follower Description** (See figure 6-1) In the dc voltage follower, the signal output $V_{OUT}$ is identical to the signal input $V_{IN}$:

$$V_{OUT} = V_{IN}$$

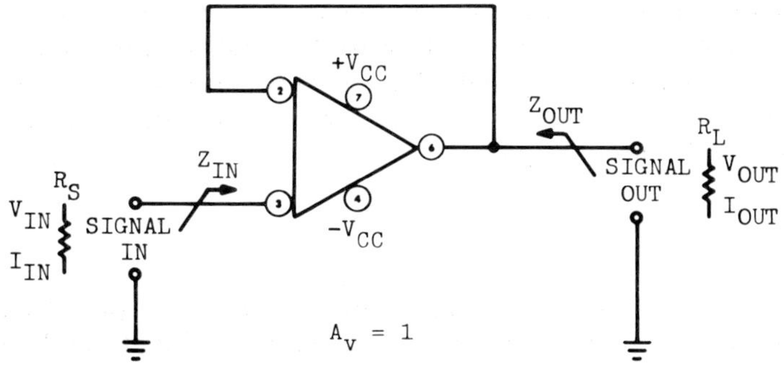


Figure 6-1 Voltage Follower (DC)

The applied voltage difference ($V_A$) in an open-loop circuit is a function of the output voltage and the open-loop gain $A_{ol}$:

$$V_A = \frac{V_{OUT}}{A_{ol}}$$

In order to change this difference voltage $V_A$, we must apply an input signal. Therefore, summing-point theory tells us that

$$V_A + V_{IN} = V_{OUT}$$

or

$$\frac{V_{OUT}}{A_{ol}} + V_{IN} = V_{OUT}$$

Remember that $V_A$ is always approaching zero.

The equation is easily established, since the circuit is considered to have open-loop gain ($A_{ol}$) without a feedback resistance. Note that 100 percent of the signal output is fed back to the inverting input. If we apply a signal to the noninverting input, the amplifier immediately drives the inverting input to approximately zero. Since the amplifier reacts to the difference voltage ($V_A$) applied between the input terminals, the output will remain at that value (zero) until the input voltage ($V_{IN}$) is changed. In equation terms:

$$V_{IN} + V_A = V_{OUT}$$

Since $V_A$ is always approaching zero, then

$$V_{IN} = V_{OUT}$$

With this circuit, the input impedance $Z_{IN}$ is infinite and the output impedance $Z_{OUT}$ is considered to be zero. As a buffer amplifier, the circuit is efficient beyond any demands. The signal is placed into the noninverting input. Therefore, there is no phase inversion. The output is essentially the same as the input. Its gain is

$$A_v = 1$$

Power, as shown in the figure 6-1, is applied to pins 4 and 7. A jumper wire is required between output pin 6 and input pin 2. For best operation, voltage outputs should not exceed 80 percent of power supply levels. Null offset should be adjusted prior to use.

## DC VOLTAGE FOLLOWER—DESIGN PROCEDURES
(See figure 6-1)

1. Select an operational amplifier that will provide the acceptable output.
2. Choose a dc power supply that will be compatible with the op amp.
3. Determine the source resistance $R_S$.
4. Determine the load resistance $R_L$.
5. Connect the circuit (see figure 6-1) and the power supply on a breadboard.
6. Adjust offset null to zero the op amp, if necessary.
7. Apply a dc signal to the noninverting input.
8. Monitor the output with a dc voltmeter. Results will be a unity gain $A_v = 1$.

## DC VOLTAGE FOLLOWER—DESIGN EXAMPLE
(See figure 6-1)

**Given:**

| | |
|---|---|
| $A_v = 1$ | $V_{IN} = 10$ vdc |
| $R_L = 2$ K | $V_{OUT} = 10$ vdc |
| $R_S = 5$ K | $Z_{IN} = 1$ MΩ |

**Procedures:**

1. Choose $\mu$A741 op amp.
2. Choose $\pm$vdc power supply.
3. Use 5 K for $R_S$.
4. Use 2 K for $R_L$.
5. Connect circuit and power supply to breadboard.
6. Adjust offset null.
7. Apply 10 vdc to noninverting input.
8. Monitor 10 vdc output with dc voltmeter.

AC Voltage Follower—Description (See figure 6-2) In the ac voltage follower circuit, the signal output $V_{OUT}$ is identical to the signal input $V_{IN}$:

$$V_{OUT} = V_{IN}$$

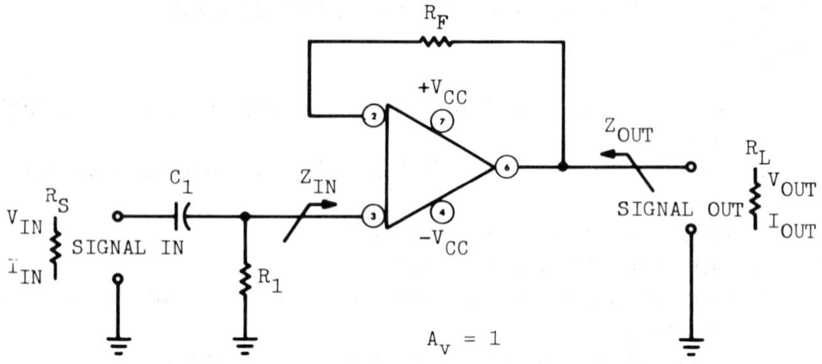


Figure 6-2 Voltage Follower (AC)

The input impedance is established by the resistor $R_1$, since $R_1$ is essentially in parallel with the open-loop impedance $Z_{IN}$. $Z_{IN}$ is, of course, much greater than $R_1$. (Note that some op amps have a $Z_{IN}$ of 15 MΩ and greater.) Therefore the choice of $R_1$ will make this external resistance the new input impedance that the source voltage sees. Remember that the total resistance of parallel legs is always smaller than the smallest resistor.

The output impedance $Z_{OUT}$ is very low and said to be zero.

Frequency range is variable and is largely dependent on the op amp chosen. The device chosen should, of course, meet frequency needs.

Power, as shown in the figure, is applied to pins 4 and 7. In figure 6-2 the $R_1$ on the noninverting input and $R_F$ on the feedback loop must be equal and much greater than the source resistance. The capacitor C1 on the input is chosen to have typical coupling characteristics. Its size is dependent on the frequency range desired.

If we apply a signal across the source resistance, it is coupled into the noninverting input. The amplifier immediately drives the inverting input to approximately zero. Since the amplifier reacts to the difference voltage $V_A$ applied between the input terminals, the output will remain at that value (zero) until the input voltage $V_{IN}$ is changed. In equation terms:

$$V_{IN} + V_A = V_{OUT}$$

Since $V_A$ is always approaching zero, then

$$V_{IN} = V_{OUT}$$

The voltage gain is therefore unity.

$$A_v = 1$$

## AC VOLTAGE FOLLOWER—DESIGN PROCEDURES

1. Select an operational amplifier that will provide the acceptable output.
2. Choose a dc power supply that will be compatible to the op amp.
3. Determine the source resistance $R_S$.
4. Choose $R_1$ (rule of thumb—$R_1 = 50\ R_S$) for minimum signal loading.
5. Select $R_F = R_1$.
6. Calculate current in: $$I_{IN} = \frac{V_{IN}}{R_1}$$
7. Determine the load resistance $R_L$.
8. Calculate load current: $$I_{OUT} = \frac{V_{OUT}}{R_L}$$
9. Calculate and choose a capacitor C1 at a low-frequency breakpoint of 10 Hz: $$C1 = \frac{1}{2\pi F(R_S + R_1)}$$
10. Connect the circuit (see figure 6-2) and the power supply on a breadboard.
11. Adjust offset null to zero the op amp, if necessary.
12. Apply an ac signal to the noninverting input.
13. Monitor the output with an ac voltmeter. Results shall be a unity gain $A_v = 1$.

## AC VOLTAGE FOLLOWER—DESIGN EXAMPLE

**Given:**

$$A_v = 1 \qquad V_{IN} = 9 \text{ vac peak}$$

$$R_L = 2\text{ K} \qquad V_{OUT} = 9 \text{ vac peak}$$

$$R_S = 1\text{ K} \qquad Z_{IN} = 1\ \Omega$$

**Procedures:**

1. Choose $\mu$A741 op amp.
2. Choose $\pm 15$ vdc power supply.
3. Use 1 K for $R_S$.
4. Choose $R_1 = 50\ R_S = 50 \times 1\ \text{K} = 50\ \text{K}$.
5. Choose $R_F = R_1 = 50\ \text{K}$.
6. Calculate $I_{\text{IN}} = \dfrac{V_{\text{IN}}}{R_1} = \dfrac{9\ \text{vac}}{50\ \text{K}} = 0.18\ \text{mA}$.
7. Use 2 K for $R_L$.
8. Calculate $I_{\text{OUT}} = \dfrac{V_{\text{OUT}}}{R_L} = \dfrac{9\ \text{vac}}{2\ \text{K}} = 4.5\ \text{mA}$.
9. Calculate $\text{C1} = \dfrac{1}{2\pi F(R_S + R_1)}$

$$= \frac{1}{6.28(10)(1\ \text{K} + 50\ \text{K})} \cong 0.31\ \mu\text{F}.$$

10. Connect circuit and power supply to breadboard.
11. Adjust offset null.
12. Apply 9 vac peak signal to noninverting input.
13. Monitor 9 vac peak output.

## INVERTING AMPLIFIER

Inverting amplifiers, using operational amplifiers, are of two basic designs: dc and ac. These amplifiers, of both types, provide high voltage gain. They are very versatile and are used in automatic control systems, sound systems, communication systems, and instrumentation. Because of the high voltage gains, the devices are usually protected by a current limiter. Inputs to the amplifiers should typically not exceed supply voltages.

DC Inverting Amplifier Description (See figure 6-3) The dc inverting amplifier using an op amp consists of the op amp, a feedback resistance $R_F$, and a metering resistance $R_M$.

Once again the author would like to remind the reader of several of the basics of op amps before proceeding with the design:

1. No current flows into the amplifier.
2. The input summing points are at virtual ground.

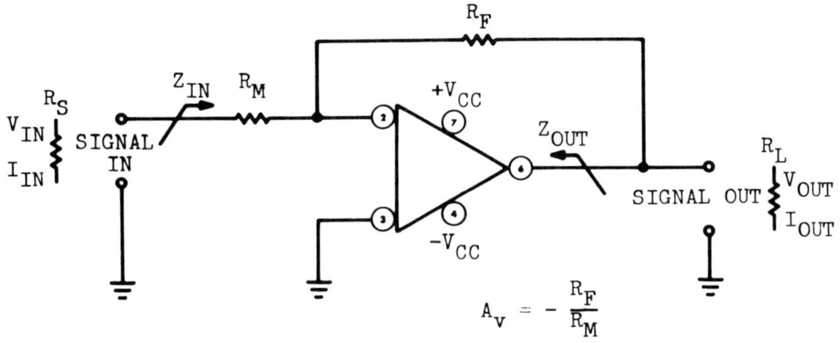


Figure 6-3 Inverting DC Amplifier

3. The difference voltage between the inputs, inverting and noninverting, is zero (0).
4. With a feedback attached (closed-loop), the negative input signal will be driven to the positive reference input.

The input impedance $Z_{IN}$ of the open-loop amplifier is infinite. However, when an external resistance is placed between the source and the summing point, such as $R_M$ in the illustration, that resistance becomes the input impedance of the circuit. Therefore:

$$R_M = Z_{IN}$$

Let's consider current flow external to the op amp. Since the summing points (terminals 2 and 3) are at virtual ground, the input current is essentially

$$I_{IN} = \frac{V_{IN} - V_A}{R_M}$$

where $V_A$ = voltage differential between the two inputs

$R_M$ = metering resistance

The current through the feedback resistance is calculated in the following manner:

$$I_{RF} = \frac{V_{OUT} - V_A}{R_F}$$

where $I_{RF}$ = feedback current

It was previously established and can be a rule now, that $I_{IN} = I_{RF}$. Therefore

$$\frac{V_{IN} - V_A}{R_M} = -\frac{V_{OUT} - V_A}{R_F}$$

Remember that the minus sign simply signifies the inverted output. Since $V_A$ = zero (0)

also
$$\frac{V_{IN}}{R_M} = -\frac{V_{OUT}}{R_F}$$

and
$$\frac{V_{OUT}}{V_{IN}} = -\frac{R_F}{R_M} = A_v$$

The voltage gain of this amplifier is then

$$A_v = -\frac{R_F}{R_M}$$

In order to match impedance in a discrete amplifier, the typical situation was to ensure that output impedance of the first stage or source was equal to or less than the second stage input impedance. Input impedance of an inverting dc amplifier is not infinite but typically around forty or fifty times the source impedance, as determined by $R_M$. This resistance decision is not set in cement. It can vary as long as input impedance is much greater than the source resistance. Output impedances are very low (25 to 50 ohms, typically) and therefore can be ignored in design.

Low output impedances are generally attributed to complementary common collector (emitter-follower) transistor configurations in the output of the amplifier. In figure 6-3 the $R_M$ resistor is chosen to be forty or fifty times the source resistor $R_S$. The resistor $R_M$ is placed between the inverting input pin 2 and the input signal. Feedback is arranged between output pin 6 and inverting pin 2. Power, as shown in figure 6-3, is applied to pins 4 and 7. A null-adjust potentiometer is placed between pins 1 and 5 to adjust the output to zero when no signal is present.

If one of the power supplies becomes disconnected or open, a large voltage may appear at output pin 6. This, of course, is dependent on the device used. In the event the device does not have internal protection, the designer should consider disabling it in the event of loss of input power.

## DC INVERTING AMPLIFIER—DESIGN PROCEDURES

1. Select an operational amplifier that will provide the acceptable output.
2. Choose a dc power supply that will be compatible with the op amp.
3. Determine the source resistance $R_S$.
4. Choose $R_M$ (rule of thumb—$R_M = 50\ R_S$) for minimum loading.
5. Calculate voltage gain required:

$$A_v = \frac{V_{\text{OUT}}}{V_{\text{IN}}} = -\frac{R_F}{R_M}$$

6. Calculate feedback resistance:

$$R_F = -A_v R_M \qquad \left(\text{since } A_v = -\frac{R_F}{R_M}\right)$$

7. Calculate current in:

$$I_{\text{IN}} = \frac{V_{\text{IN}}}{R_M}$$

8. Calculate load current:

$$I_{\text{OUT}} = \frac{V_{\text{OUT}}}{R_L}$$

9. Connect the circuit (see figure 6-3) and the power supply on a breadboard.
10. Adjust offset null to zero the op amp, if necessary.
11. Apply a dc signal to the inverting input.
12. Monitor the output with a dc voltmeter. Results shall be the calculated gain $A_v$.

## DC INVERTING AMPLIFIER—DESIGN EXAMPLE

**Given:**

$$R_L = 2\text{ K} \qquad V_{\text{IN}} = 120\text{ mvdc}$$

$$R_S = 100\ \Omega \qquad V_{\text{OUT}} = 12\text{ vdc}$$

**Procedures:**

1. Choose $\mu$A741 op amp.
2. Choose $\pm 15$ vdc power supply.
3. Use 100 $\Omega$ for $R_S$.
4. Choose $R_M = 50\ R_S = 50 \times 100 = 5$ K.
5. Calculate $A_v = \dfrac{V_{OUT}}{V_{IN}} = \dfrac{12 \text{ vdc}}{120 \text{ mvdc}} = 100.$
6. Calculate $R_F = -A_v R_M = 100 \times 5 \text{ K} = 500$ K.
7. Calculate $I_{IN} = \dfrac{V_{IN}}{R_M} = \dfrac{120 \text{ mvdc}}{5 \text{ K}} = 24\ \mu\text{A}.$
8. Calculate $I_{OUT} = \dfrac{V_{OUT}}{R_L} = \dfrac{12 \text{ vdc}}{2 \text{ K}} = 6$ mA.
9. Connect circuit and power supply to breadboard.
10. Adjust offset null.
11. Apply 120 mvdc to inverting input.
12. Monitor $-12$ vdc output.

AC Inverting Amplifier Description (See figure 6-4) In the inverting ac amplifier, large signal voltage gain may approach values such as 100,000. Input impedance should be much greater than source resistance (for example, a 100-ohm source resistance and 5 K ohms input impedance). Output impedances are very low, with 25 to 50 ohms typical, and therefore can be ignored in design.

Low-output impedances are generally attributed to a common collector (emitter-follower) transistor configuration in the output of the amplifier. In figure 6-4 the resistor $R_M$ should be much greater than the source resistance. $R_M$ resistor can be said to be equal to input impedance $Z_{IN}$, since pin 2 is a virtual ground because of pin 3 ground connection. Resistor $R_M$ is installed between inverting input pin 2 and the input signal. Feedback in this configuration is arranged between output pin 6 and inverting pin 2. Power as shown in the figure is applied to pins 4 and 7.

If one of the power supplies becomes disconnected or open, a large voltage may appear at output pin 6. This, of course, is dependent on the device used. In the event the device does not have internal protection, the designer should consider disabling it in the event of loss of input power. Voltage gain is:

$$A_v = -\frac{R_F}{R_M}$$

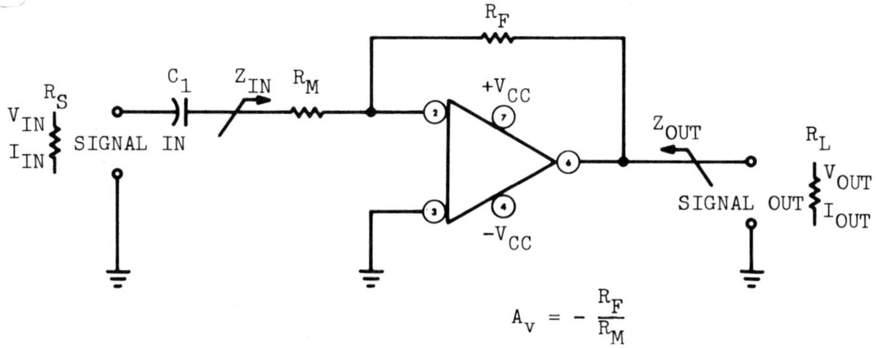


Figure 6-4 Inverting AC Amplifier

The input capacitor is chosen as a function of frequency and input reactance. This design is good for frequencies up to 10 kHz if a μA741 amplifier is used. At that frequency, the signal will roll off at 20 dB per decade. For other designs, manufacturers' specifications should be consulted, because frequency response varies with the device used and gain requirements.

All dc analysis and calculations of gain from the dc inverting amplifier design are applicable to the ac inverting amplifier.

## AC INVERTING AMPLIFIER—DESIGN PROCEDURES (See figure 6-4)

1. Select an operational amplifier that will provide the acceptable output.
2. Choose a dc power supply that will be compatible with the op amp.
3. Determine the source resistance $R_S$.
4. Choose $R_M$ (rule of thumb—$R_M = 50\ R_S$) for minimum loading.
5. Calculate voltage gain required:

$$A_v = \frac{V_{OUT}}{V_{IN}} = -\frac{R_F}{R_M}$$

6. Calculate feedback resistance:

$$R_F = -A_v R_M \qquad \left(\text{since } A_v = \frac{-R_F}{R_M}\right)$$

7. Calculate current in:

$$I_{IN} = \frac{V_{IN}}{R_M}$$

8. Calculate current out:

$$I_{OUT} = \frac{V_{OUT}}{R_L}$$

9. Calculate capacitor C1 at a break frequency of 10 hertz:

$$C1 = \frac{1}{2\pi F(R_S + R_M)}$$

10. Connect the circuit (see figure 6-4) and the power supply on a breadboard.
11. Adjust offset null to zero the op amp, if necessary.
12. Apply an ac signal to the inverting input.
13. Monitor the output with an ac voltmeter. Results shall be the calculated gain $A_v$.

## AC INVERTING AMPLIFIER—DESIGN EXAMPLE (See figure 6-4)

**Given:**

$R_L = 2$ K $\quad V_{IN} = 90$ mvac peak at 1,000 Hz

$R_S = 100\ \Omega \quad V_{OUT} = 9$ vac peak

**Procedures:**

1. Choose $\mu$A741.
2. Choose $\pm 15$ vdc power supply.
3. Use 100 Ω for $R_S$.
4. Choose $R_M = 50\ R_S = 50 \times 100 = 5$ K.
5. Calculate $A_v = \frac{V_{OUT}}{V_{IN}} = \frac{9 \text{ vac}}{90 \text{ mvac}} = 100.$
6. Calculate $R_F = -A_v R_M = 100 \times 5 \text{ K} = 500$ K.
7. Calculate $I_{IN} = \frac{V_{IN}}{R_M} = \frac{90 \text{ mvac}}{5 \text{ K}} = 18\ \mu\text{A}.$

8. Calculate $I_{\text{OUT}} = \dfrac{V_{\text{OUT}}}{R_L} = \dfrac{9 \text{ vac}}{2 \text{ K}} = 4.5 \text{ mA}.$

9. Calculate C1 $= \dfrac{1}{2\pi F(R_S + R_M)}$

$$= \frac{1}{6.28(10)(5 \text{ K})} = 3.1\ \mu\text{F}.$$

10. Connect circuit and power supply to breadboard.
11. Adjust offset null.
12. Apply 9 mvac peak signal at 1,000 Hz to inverting input.
13. Monitor the output with an ac voltmeter. Results shall be the calculated gain $A_v$.

## NONINVERTING AMPLIFIER

Noninverting amplifiers, using operational amplifiers, provide high voltage gain. Since this is the case, the output is usually protected by a current limiter. Inputs to the noninverting amplifier should typically not exceed the supply voltage. Large signal voltage gain may approach values such as 100,000.

Noninverting amplifiers, like inverting amplifiers, are used in automatic control systems, sound systems, communication systems, and instrumentation. Also, like inverting amplifiers, the noninverting amplifiers are basically of two types: dc and ac.

**DC Noninverting Amplifier Description** (See figure 6-5) The dc noninverting amplifier using an op amp consists of the op amp, a feedback resistance $R_F$, and a metering resistance $R_M$. The input impedance at pin 3 is infinitive.

To develop a gain equation, the current flow external to the op amp must be considered. Since the summing points (terminals 2 and 3) are at virtual ground, the current through $R_M$ is essentially as follows:

$$I_{RM} = \frac{V_{RM}}{R_M}$$

The current flow through the feedback resistance is

$$I_{RF} = \frac{V_{RF}}{R_F}$$

You may remember that $I_{RM} = I_{RF}$ from the previous discussion.

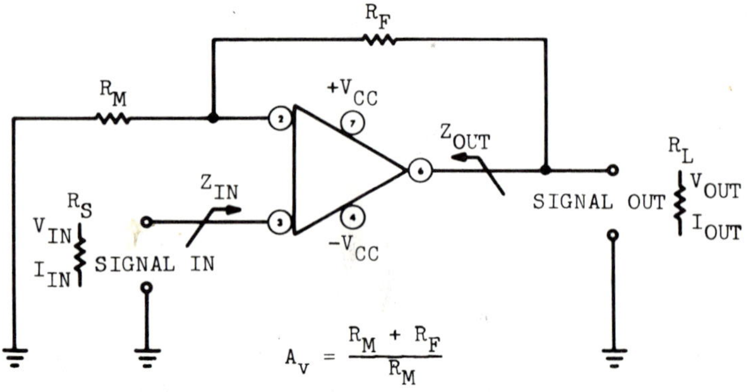


Figure 6-5 Noninverting DC Amplifier

Another previous discussion on the inverting amplifier proved that the voltage at the inverting terminal is $V_{IN} + V_A$, where $V_A$ was the applied voltage difference between the terminals 2 and 3.

The voltage at the inverting terminal is $V_{IN} + V_A$, therefore

$$I_{RM} = \frac{V_{IN} + V_A}{R_M}$$

The voltage at the feedback resistance is $V_{OUT} - (V_{IN} + V_A)$, therefore

$$I_{RF} = \frac{V_{OUT} - (V_{IN} + V_A)}{R_F}$$

Since $I_{RM} = I_{RF}$, then

$$\frac{V_{IN} + V_A}{R_M} = \frac{V_{OUT} - (V_{IN} + V_A)}{R_F}$$

We have previously assumed that $V_A$ = zero. We can therefore rewrite the equation

$$\frac{V_{IN}}{R_M} = \frac{V_{OUT} - V_{IN}}{R_F}$$

Rewriting this formula provides

$$V_{IN}\,R_F = R_M\,V_{OUT} - R_M\,V_{IN}$$

$$V_{IN}\,R_F + V_{IN}\,R_M = V_{OUT}\,R_M$$

$$\frac{V_{IN}\,(R_F + R_M)}{R_M} = V_{OUT}$$

and

$$A_v = \frac{V_{OUT}}{V_{IN}} = \frac{R_F + R_M}{R_M}$$

or

$$A_v = \frac{R_F}{R_M} + 1$$

The noninverting amplifier has the advantage of near infinity input the same as the voltage follower. It therefore is used as a buffer for impedance matching. Its output impedance can be said to be zero.

Low-output impedances are generally attributed to a common collector (emitter-follower) transistor configuration in the output of the amplifier. In figure 6-5, the $R_M$ resistor is chosen to match the source impedance and placed between the inverting input pin 2 and ground. Feedback is arranged between output pin 6 and inverting pin 2. Power, as shown in the figure, is applied to pins 4 and 7.

If one of the power supplies becomes disconnected or open, a large voltage may appear at output pin 6. This of course, is dependent on the device used. Voltage gain is:

$$A_v = \frac{R_M + R_F}{R_M}$$

## DC NONINVERTING AMPLIFIER—DESIGN PROCEDURES (See figure 6-5)

1. Select an operational amplifier that will provide the acceptable output.
2. Choose a dc power supply that will be compatible with the op amp.
3. Determine the source resistance $R_S$.
4. Choose metering resistance $R_M$ to equal source resistance $R_S$.
5. Calculate voltage gain required:

$$A_v = \frac{V_{OUT}}{V_{IN}} = \frac{R_M + R_F}{R_M}$$

6. Calculate feedback resistance:

$$R_F = A_v R_M - R_M$$

7. Calculate current in:

$$I_N = \frac{V_{IN}}{Z_{IN}}$$

8. Calculate current out:

$$I_{OUT} = \frac{V_{OUT}}{R_L}$$

9. Connect the circuit (see figure 6-5) and the power supply on a breadboard.
10. Adjust offset null to zero the op amp, if necessary.
11. Apply a dc signal to the inverting input.
12. Monitor the output with a dc voltmeter. Results shall be the calculated gain $A_v$.

## DC NONINVERTING AMPLIFIER—DESIGN EXAMPLE (See figure 6-5)

**Given:**

$$R_L = 2\text{ K} \qquad V_{IN} = 120\text{ mvdc}$$

$$R_S = 5\text{ K} \qquad V_{OUT} = 12\text{ vdc}$$

$$Z_{IN} = 1\text{ M}\Omega$$

**Procedures:**

1. Choose $\mu$A741 op amp.
2. Choose $\pm 15$ vdc power supply.
3. Use 5 K for $R_S$.
4. Choose $R_M = 5$ K.
5. Calculate $A_v = \dfrac{V_{OUT}}{V_{IN}} = \dfrac{12\text{ vdc}}{120\text{ mvdc}} = 100.$
6. Calculate $R_F = A_v R_M - R_M = 100(5\ K) - 5\text{ K} = 495\text{ K}.$
7. Calculate $I_N = \dfrac{V_{IN}}{Z_{IN}} = \dfrac{120\text{ mvdc}}{1\text{ M}\Omega} = .12\ \mu\text{A}.$

8. Calculate $I_{OUT} = \dfrac{V_{OUT}}{R_L} = \dfrac{12\text{ vdc}}{2\text{ K}} = 6\text{ mA}$.
9. Connect circuit and power supply to breadboard.
10. Adjust offset null.
11. Apply 120 mvdc to noninverting input.
12. Monitor 12 vdc output.

**The Adjustable Offset Noninverting Amplifier (See figure 6-6)** The adjustable offset noninverting dc amplifier is identical to the standard noninverting dc amplifier in operation and is used interchangeably with it. The null-offset feature allows the designer to get symmetric amplifier output swing for inputs that are offset.

The difference between this design and the standard noninverting dc amplifier is the addition of a variable resistor on the inverting input, along with a voltage divider.

The variable resistor $R_1$ can be any value, with 50 K ohms typical. $R_3$ and $R_4$ feedback voltage divider should be equal to the source resistance $R_S$. $R_2$ is typically 100 K ohms. Voltage gain is:

$$A_v = 1 + \frac{R_F}{R_3 + R_4}$$

## ADJUSTABLE OFFSET NONINVERTING DC AMPLIFIER—DESIGN PROCEDURES (See figure 6-6)

1. Select an operational amplifier that will provide the acceptable output.
2. Choose a dc power supply that will be compatible with the op amp.
3. Determine the source resistance $R_S$.
4. Choose a potentiometer $R_1$ (typically 50 K).
5. Choose divider resistance $R_2$ (typically 100 K).
6. Calculate $R_3$ for desired bias range.

$$\text{where bias range} = \pm V_{CC}\frac{R_3}{R_1}$$

then

$$R_3 = \frac{\text{bias range } R_2}{\pm V_{CC}}$$

7. Calculate divider resistance $R_4$:

$$R_4 = R_S - R_3$$

8. Calculate voltage gain required:

$$A_v = \frac{V_{\text{OUT}}}{V_{\text{IN}}}$$

9. Calculate feedback resistor $R_F$:

$$R_F = (R_3 + R_4)(A_v - 1)$$

10. Connect the circuit (see figure 6-6) and the power supply to a breadboard.
11. Remove $\pm V_{CC}$ reference, null $V_{\text{IN}}$ signal, and adjust offset null potentiometer to zero the op amp at the output.
12. Apply a $\pm V_{CC}$ dc reference and adjust potentiometer for maximum positive output.
13. Monitor output for +vdc using dc voltmeter.
14. Apply +mvdc at $V_{\text{IN}}$.
15. Monitor output for +vdc using dc voltmeter.
16. Apply −mvdc at $V_{\text{IN}}$.
17. Monitor output for −vdc equal to +vdc in step 15.
18. Adjust bias potentiometer for maximum negative output.
19. Monitor output for −vdc using dc voltmeter.

## ADJUSTABLE OFFSET NONINVERTING DC AMPLIFIER—DESIGN EXAMPLE (See figure 6-6)

**Given:**

bias range = ±30 mvdc $\qquad R_L = 2$ K

$V_{\text{IN}} = 90$ mvdc $\qquad R_S = 1$ K

$V_{\text{OUT}} = 12$ vdc ($V_{\text{IN}}$ + max bias)

**Procedures:**

1. Choose $\mu$A741 op amp.
2. Choose ±15 vdc power supply.
3. Select 1 K for $R_S$.
4. Choose $R_1 = 50$ K typical.
5. Choose $R_2 = 100$ K typical.
6. Calculate bias range resistance $R_3$:

$$R_3 = \frac{\text{bias range } R_2}{\pm V_{CC}}$$

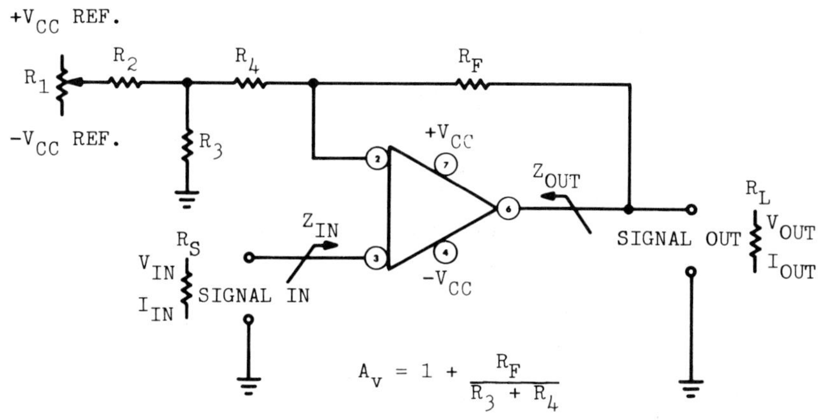


Figure 6-6 Adjustable Offset Noninverting DC Amplifier

$$R_3 = \frac{30 \text{ mvdc} \times 100 \text{ K}}{15}$$

$$R_3 = 200\ \Omega$$

7. Calculate $R_4$ :

$$R_4 = R_S - R_3$$

$$R_4 = 1 \text{ K} - 200\ \Omega = 800\ \Omega$$

8. Calculate voltage gain required:

$$A_v = \frac{V_{\text{OUT}}}{V_{\text{IN}}}$$

$$A_v = \frac{12 \text{ vdc}}{(90 + 30) \text{ mvdc}} = 100$$

9. Calculate feedback resistor $R_F$ :

$$R_F = (R_3 + R_4)(A_v - 1)$$

$$R_F = (200\ \Omega + 800\ \Omega)(100 - 1) = 99 \text{ K}$$

10. Connect circuit and power supply to a breadboard.
11. Adjust offset potentiometer to a null at the output.
12. Apply $\pm V_{CC}$ reference and adjust bias potentiometer for maximum positive output.

13. Monitor output for 3 vdc using dc voltmeter.
14. Apply positive 90 mvdc at $V_{IN}$.
15. Output should indicate +12 vdc on dc voltmeter.
16. Apply negative 90 mvdc at $V_{IN}$.
17. Output should indicate −6 vdc on dc voltmeter.
18. Adjust bias potentiometer for maximum negative output.
19. Monitor output for −12 vdc on dc voltmeter.
    **Note:** With this feature, null point can be set in range from +3 vdc to −3 vdc at output (see figure 6-7).

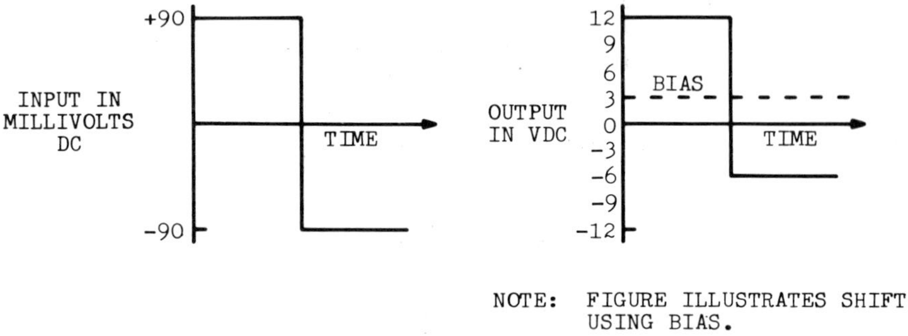


Figure 6-7 Input Versus Output Volts DC Adjustable Offset Noninverting Amplifier

AC Noninverting Amplifier Description (See figure 6-8) Components for the ac noninverting amplifier are the same as for the dc noninverting amplifier except for the added ac coupling capacitor.

Input impedance of a noninverting ac amplifier is typically 5 K to 10 K ohms, as determined by $R_N$. Output impedances are very low (25 to 50 ohms typically) and therefore can be ignored in design.

Low output impedances are generally attributed to complementary common collector (emitter-follower) transistor configurations in the output of the amplifier. In figure 6-8, the $R_M$ resistor is placed between the inverting input 2 and ground. Feedback is arranged between output pin 6 and inverting input pin 2. The resistor $R_N$ is placed between the noninverting input pin 3 and ground. The capacitor $C_1$ is placed between the resistor $R_N$ and the input signal. The coupling capacitor $C_1$ is chosen to pass the lowest frequency to be encountered without attenuating. Power, as shown in figure 6-8, is applied to pins 4 and 7. A null-adjust potentiometer is placed between pin 2 and 5 to adjust the output to zero when no signal is present.

If one of the power supplies becomes disconnected or open, a large

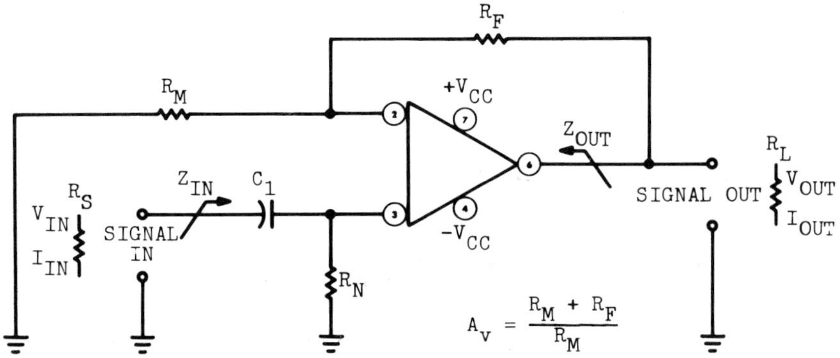


Figure 6-8 Noninverting AC Amplifier

voltage may appear at output pin 6. This, of course, is dependent on the device used. Voltage gain is:

$$A_v = \frac{R_M + R_F}{R_M}$$

All dc analysis and calculations of gain for the dc noninverting amplifier are applicable to the ac noninverting amplifier.

## AC NONINVERTING AMPLIFIER—DESIGN PROCEDURES (See figure 6-8)

1. Select an operational amplifier that will provide the acceptable output.
2. Choose a dc power supply that will be compatible with the op amp.
3. Determine the source resistance $R_S$.
4. Choose metering resistance $R_M$ to be equal to $R_S$.
5. Choose noninverting resistance input $R_N$ to be equal to $R_S$.
6. Calculate voltage gain required:

$$A_v = \frac{V_{\text{OUT}}}{V_{\text{IN}}} = \frac{R_M + R_F}{R_M}$$

7. Calculate current in:

$$I_{\text{IN}} = \frac{V_{\text{IN}}}{R_N}$$

8. Calculate current out:

$$I_{OUT} = \frac{V_{OUT}}{R_L}$$

9. Calculate capacitor $C_1$ at a break frequency of 10 hertz:

$$C_1 = \frac{1}{2\pi F(R_S + R_N)}$$

10. Connect the circuit (see figure 6-8) and the power supply on a breadboard.
11. Adjust offset null to zero the op amp, if necessary.
12. Apply an ac signal to the noninverting input.
13. Monitor the output with an ac voltmeter. Results shall be the calculated gain $A_v$.

## AC NONINVERTING AMPLIFIER—DESIGN EXAMPLE (See figure 6-8)

**Given:**

$R_L = 2$ K $\qquad V_{IN} = 90$ mvac at 1,000 Hz

$R_S = 5$ K $\qquad V_{OUT} = 9$ vac peak

**Procedures:**

1. Choose $\mu$A741.
2. Choose $\pm 15$ vdc power supply.
3. Use 5 K for $R_S$.
4. Choose $R_M = 5$ K.
5. Choose $R_N = 5$ K.
6. Calculate $A_v = \frac{V_{OUT}}{V_{IN}} = \frac{9 \text{ vac}}{90 \text{ mvac}} = 100.$
7. Calculate $I_{IN} = \frac{V_{IN}}{R_N} = \frac{90 \text{ mvac}}{5 \text{ K}} = 18\ \mu\text{A}.$
8. Calculate $I_{OUT} = \frac{V_{OUT}}{R_L} = \frac{9 \text{ vac}}{2 \text{ K}} = 4.5$ mA.
9. Calculate $C_1 = \frac{1}{2\pi F(R_S + R_M)} = \frac{1}{6.28(10)(5 \text{ K} + 5 \text{ K})} = 1.6\ \mu\text{F}.$

10. Connect circuit and power supply to breadboard.
11. Adjust offset null.
12. Apply 90 mvac peak signal at 1,000 Hz to the noninverting input.
13. Monitor the output with an ac voltmeter. Results shall be 9 vac peak at 1,000 Hz.

## DIFFERENTIAL AMPLIFIER

Differential dc amplifiers, using operational amplifiers, are of two basic types: dc and ac. Both types provide low voltage gain and high current gain. These amplifiers are used in applications where input differences must be amplified, and their primary use is in servo control systems and instrumentation.

**DC Differential Amplifier Description** (See figure 6-9) The dc differential amplifier using an op amp consists of the op amp, a feedback resistance $R_F$, a metering resistance $R_M$, and the voltage divider resistances $R_N$ and $R_V$ connected at the noninverting input.

To develop a gain equation for the differential amplifier, one must consider several basic considerations:

1. No current flows into the amplifier. The current $I_{RM}$ and the current $I_{RF}$ are equal. Likewise, the current $I_{RN}$ and the current $I_{RV}$ are equal.
2. The ideal voltage situation is that the inputs $V_{IN1}$ and $V_{IN2}$ track each other. The difference input voltage is zero and the output difference voltage is zero.

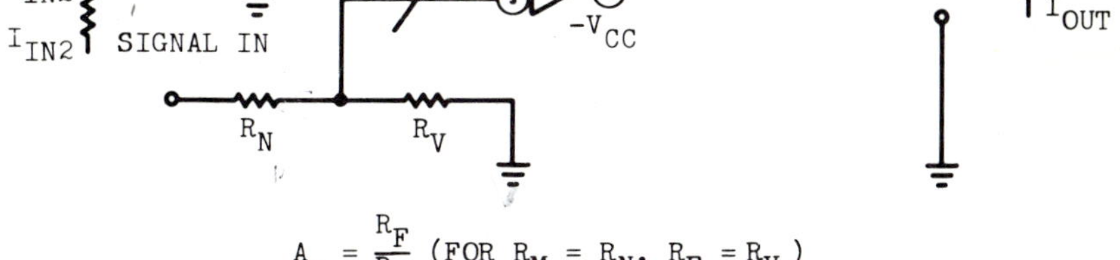


$$A_V = \frac{R_F}{R_M} \text{ (FOR } R_M = R_N,\ R_F = R_V\text{)}$$

Figure 6-9 Differential DC Amplifier

3. If the inverting and noninverting inputs are equal they will cancel each other out inside the amplifier.
4. The summing points (terminals 2 and 3) are at virtual ground.
5. The resistances $R_M$ and $R_N$ are equal. The resistances $R_F$ and $R_V$ are also equal.

For explanation purposes, let's assume that the voltage input $V_{IN2}$ is at zero. The voltage output with $V_{IN1}$ only applied at the inverting terminal is as follows:

$$V_{OUT} = A_v V_{IN}$$

or

$$V_{OUT} = -\frac{R_F}{R_M} V_{IN}$$

where

$$A_v = -\frac{R_F}{R_M}$$

A similar assumption can be made with the voltage input with $V_{IN1}$ at zero. The voltage output with $V_{IN2}$ only applied at the noninverting terminal is as follows:

$$V_{OUT} = A_v \cdot V_{IN}$$

or

$$V_{OUT} = \frac{R_F + R_M}{R_M} V_{IN}$$

where

$$A_v = \frac{R_F + R_M}{R_M}$$

The amplifier will act like a noninverting amplifier with the voltage divider $R_N$ and $R_V$ supplying the input voltage to terminal 3:

$$V_{IN2} \frac{R_V}{R_N + R_V} = \text{volts at terminal 3}$$

If (and when) both input signals $V_{IN1}$ and $V_{IN2}$ are applied simultaneously, the signal $V_{IN1}$ causes the output to drive terminal 2 voltage to equal the voltage at terminal 3.

Terminal 3 voltage at this point is the value

$$V_{IN2} \frac{R_V}{R_N + R_V}$$

instead of ground (or zero).

The differential output voltage equation can be examined with these input values:

$$V_{OUT} = V_{IN2}\left(\frac{R_V}{R_N + R_V}\right)\left(\frac{R_M + R_F}{R_M}\right) - V_{IN1}\,\frac{R_F}{R_M}$$

If the equalities $R_M = R_N$ and $R_F = R_V$, the equation can be reduced to

$$V_{OUT} = V_{IN2}\,\frac{R_F}{R_M} - V_{IN1}\,\frac{R_F}{R_M}$$

and further,

$$V_{OUT} = \frac{R_F}{R_M}\,(V_{IN2} - V_{IN1})$$

This equation, as you can see, is similar to the dc inverting amplifier gain formula.

We have described a great deal with numbers up to this point. Let's now try to unscramble this differential amplifier function with a few selected words and numbers.

We can assume that the amplifier has a voltage gain of 2. Assume also that +2 V are placed on each input. The output would be zero.

Suppose +1 V is placed on the inverting input and −1 V is placed on the noninverting input. The differential voltage output is as follows:

$$V_{OUT} = A_v(V_{IN2} - V_{IN1})$$

$$V_{OUT} = 2[-1 - (+1)]$$

$$V_{OUT} = 2(-2) = -4\text{ V}$$

If +2 V is placed on the inverting input and +4 V is placed on the noninverting input, the differential voltage output is as follows:

$$V_{OUT} = A_v(V_{IN2} - V_{IN1})$$

$$V_{OUT} = 2[+4 - (+2)]$$

$$V_{OUT} = 2(+2) = +4\text{ V}$$

If −2 V is placed on the inverting input and −4 volts is placed on the noninverting input, the differential output is as follows:

$$V_{OUT} = A_v(V_{IN2} - V_{IN1})$$

$$V_{OUT} = 2[-4 - (-2)]$$

$$V_{OUT} = 2(-2) = -4\text{ V}$$

Inputs to the differential dc amplifier are not critical, and can be any voltage up to 3 or 5 volts, depending on the power-supply level. Amplifier input impedances are large (near 1 megohm). Output impedances are very small (less than 25 ohms) and can be ignored in design.

In figure 6-9, $R_F$ and $R_M$ are used as feedback and inverting input voltage divider, respectively. $R_F$ feedback is placed between output pin 6 and inverting pin 2. $R_N$ and $R_V$ are the noninverting input signal voltage dividers for input terminal 3 voltage. The illustration shows the rest of the installation.

Power is applied to pins 4 and 7. Null adjustment, though not shown, is between pins 1 and 5. Voltage gain is:

$$A_v = \frac{R_F}{R_M}$$

## DC DIFFERENTIAL AMPLIFIER—DESIGN PROCEDURES (See figure 6-9)

1. Select an operational amplifier that will provide the acceptable output.
2. Choose a dc power supply that will be compatible with the op amp.
3. Determine the source resistances $R_{S1}$ and $R_{S2}$.
4. Choose $R_M$ to be large (rule of thumb—$R_M = 50\ R_{S1}$).
5. Choose $R_N$ to be large and equal to $R_M$.
6. Calculate voltage gain required:

$$A_v = \frac{V_{\text{OUT}}}{V_{\text{IN1}} - V_{\text{IN2}}} = -\frac{R_F}{R_M}$$

7. Calculate feedback resistance:

$$R_F = -A_v R_M$$

8. Choose voltage divider resistance $R_V$ to be equal to $R_F$.
9. Calculate current in:

$$I_{\text{IN1}} = \frac{V_{\text{IN1}}}{R_{S1} + R_M + \dfrac{R_F Z_{\text{IN}}}{R_F + Z_{\text{IN}}}}$$

$$I_{IN2} = \frac{-V_{IN2}}{R_{S2} + R_N + \dfrac{R_V Z_{IN}}{R_V + Z_{IN}}}$$

10. Calculate current output:

$$I_{OUT} = \frac{V_{OUT}}{R_L}$$

11. Connect the circuit and the power supply on a breadboard.
12. Adjust offset null to zero the op amp if necessary.
13. Apply the same signal at $V_{IN1}$ and $V_{IN2}$ inputs.
14. Monitor the output with a dc voltmeter.
15. Apply a positive signal at $V_{IN1}$ and a negative signal at $V_{IN2}$.
16. Monitor the output with a dc voltmeter.
17. Apply a negative signal at $V_{IN1}$ and a positive signal at $V_{IN2}$.
18. Monitor the output with a dc voltmeter.
    **Note:** The output is a function of the potential difference between the two inputs and the amplifier gain.

## DC DIFFERENTIAL AMPLIFIER—DESIGN EXAMPLE
(See figure 6-9)

**Given:**

$R_L = 2$ K $\qquad V_{IN1} = \pm 1$ V

$R_{S1} = 500\ \Omega \qquad V_{IN2} = \pm 1$ V

$R_{S2} = 500\ \Omega$

**Procedures:**

1. Choose $\mu$A741 op amp.
2. Choose $\pm 15$ vdc power supply.
3. Use 500 Ω for $R_{S1}$ and $R_{S2}$.
4. Choose $R_M = 50\ R_{S1} = 50 \times 500 = 25$ K.
5. Choose $R_N = R_M = 25$ K.
6. Calculate $A_v = \dfrac{V_{OUT}}{V_{IN1} - V_{IN2}} = \dfrac{+10 \text{ vdc}}{1 \text{ V} - (-1)} = \dfrac{10}{2} = 5.$
7. Calculate $R_F = -A_v R_M = -5(25 \text{ K}) = 125$ K.

8. Choose $R_V = R_F = 125$ K.
9. Calculate current in:

$$I_{IN1} = \frac{V_{IN1}}{R_{S1} + R_M + \dfrac{R_F Z_{IN}}{R_F + Z_{IN}}}$$

$$= \frac{1}{500\ \Omega + 25\ \text{K} + \dfrac{125\ \text{K}(1\ \text{M})}{125\ \text{K} + 1\ \text{M}}} = 7.3\ \mu\text{A}$$

$$I_{IN2} = \frac{-V_{IN2}}{R_{S2} + R_N + \dfrac{R_V Z_{IN}}{R_V + Z_{IN}}}$$

$$= \frac{1}{500\ \Omega + 25\ \text{K} + \dfrac{125\ \text{K}(1\ \text{M})}{125\ \text{K} + 1\ \text{M}}} \cong -7.3\ \mu\text{A}$$

10. Calculate current out:

$$I_{OUT} = \frac{V_{OUT}}{R_L} = \frac{10\ \text{vdc}}{2\ \text{K}} = 5\ \text{mA}$$

11. Connect circuit and power supply to breadboard.
12. Adjust offset null.
13. Apply +1 vdc at $V_{IN1}$ and $V_{IN2}$.
14. Monitor zero vdc output with dc voltmeter.
15. Apply +1 vdc at $V_{IN1}$ and −1 vdc at $V_{IN2}$.
16. Monitor −10 vdc output with dc voltmeter.
17. Apply −1 vdc at $V_{IN1}$ and +1 vdc at $V_{IN2}$.
18. Monitor 10 vdc output with dc voltmeter.

**AC Differential Amplifier Description (See figure 6-10)** Components for the ac differential amplifier are the same as for the dc differential amplifier except for the added ac coupling capacitors.

The differential ac amplifier, using an operational amplifier, has a low voltage gain (typically under 25). Amplifier input impedance is very large (in the realm of 1 megohm), while output impedance approaches zero (less than 25 ohms) and can be said to be zero.

The resistors $R_N$ and $R_V$ are voltage dividers for the noninverting input. $R_F$ feedback resistor is placed between output pin 6 and inverting input pin 2. The metering resistor $R_M$ is placed between $V_{IN1}$ and the

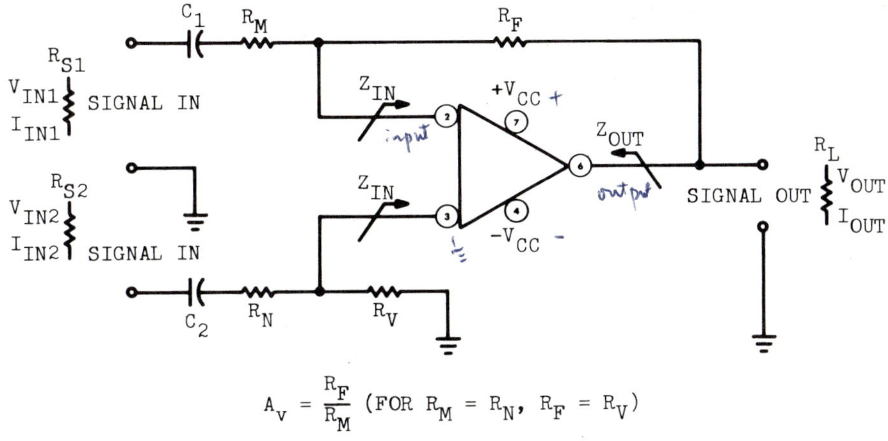


Figure 6-10 Differential AC Amplifier

inverting input pin 2. The capacitors $C_1$ and $C_2$ are coupling capacitors. Their sizes are selected to pass the lowest frequency to be encountered without attenuation.

Power is applied to pins 4 and 7, offset null adjust is between pin 1 and 5. Except for the input capacitors, the differential ac amplifier is designed in much the same manner as the differential dc amplifier. Voltage gain is:

$$A_v = \frac{R_F}{R_M}$$

## AC DIFFERENTIAL AMPLIFIER—DESIGN PROCEDURES (See figure 6-10)

1. Select an operational amplifier that will provide the acceptable output.
2. Choose a dc power supply that will be compatible with the op amp.
3. Determine the source resistances $R_{S1}$ and $R_{S2}$.
4. Choose $R_M$ to be large (rule of thumb—$R_M = 50\ R_{S1}$).
5. Choose $R_N$ to large and equal to $R_M$.
6. Calculate voltage gain required:

$$A_v = \frac{V_{\text{OUT}}}{V_{\text{IN1}} - V_{\text{IN2}}} = -\frac{R_F}{R_M}$$

7. Calculate feedback resistance $R_F = -A_v R_M$.
8. Choose voltage divider resistance $R_V$ to be equal to $R_F$.
9. Calculate current in:

$$I_{IN1} = \frac{V_{IN1}}{R_{S1} + R_M + \dfrac{R_F Z_{IN}}{R_F + Z_{IN}}}$$

$$I_{IN2} = \frac{-V_{IN2}}{R_{S2} + R_N + \dfrac{R_V Z_{IN}}{R_V + Z_{IN}}}$$

10. Calculate $C_1$ and $C_2$ capacitor sizes:

$$C_1 = \frac{1}{2\pi F\left(R_{S1} + R_M + \dfrac{R_F Z_{IN}}{R_V + Z_{IN}}\right)}$$

$$C_2 = \frac{1}{2\pi F\left(R_{S2} + R_N + \dfrac{R_V Z_{IN}}{R_V + Z_{IN}}\right)}$$

11. Calculate current output:

$$I_{OUT} = \frac{V_{OUT}}{R_L}$$

12. Connect the circuit and the power supply on a breadboard.
13. Adjust offset null to zero the amp if necessary.
14. Apply the same phase signal at $V_{IN1}$ and $V_{IN2}$ inputs.
15. Monitor the output with an ac voltmeter.
16. Apply an out-of-phase signal at $V_{IN1}$ and an in-phase signal at $V_{IN2}$.
17. Monitor the output with an ac voltmeter.
18. Apply an in-phase signal at $V_{IN1}$ and an out-of-phase signal at $V_{IN2}$.
19. Monitor the output with an ac voltmeter.
    **Note:** The output is a function of the potential difference between the two inputs and the amplifier gain.

## AC DIFFERENTIAL AMPLIFIER—DESIGN EXAMPLE
(See figure 6-10)

**Given:**

| | |
|---|---|
| $R_L = 2$ K | $V_{IN1} = 1$ vac peak at 1,000 Hz |
| $R_{S1} = 500\ \Omega$ | $V_{IN2} = -1$ vac peak at 1,000 Hz |
| $R_{S2} = 500\ \Omega$ | $V_{OUT} = -9$ vac peak |
| | $Z_{IN} = 1\ \text{M}\Omega$ |

**Procedure:**

1. Choose $\mu$A741 op amp.
2. Choose $\pm$vdc power supply.
3. Use 500 Ω for $R_{S1}$ and $R_{S2}$.
4. Calculate $R_M = 50\ R_{S1} = 25$ K.
5. Choose $R_N = R_M = 25$ K.
6. Calculate $A_v = \dfrac{V_{OUT}}{V_{IN1} - V_{IN2}}$

$$= \frac{9 \text{ vac}}{1 \text{ vac} - (-1 \text{ vac})} = \frac{9}{2} = 4.5.$$

7. Calculate $R_F = -A_v R_M = 4.5 \times 25\ \text{K} = 113$ K.
8. Choose $R_V = R_F = 113$ K.
9. Calculate $I_{IN1} = \dfrac{1}{R_{S1} + R_M + \dfrac{R_F Z_{IN}}{R_F + Z_{IN}}}$

$$= \frac{1}{500\ \Omega + 25\ \text{K} + \dfrac{113\ \text{K} \times 1\ \text{M}}{113\ \text{K} + 1\ \text{M}}}$$

$$= \frac{1}{127\ \text{K}} \cong 7.9\ \mu\text{A}$$

$$I_{IN2} = \frac{-1}{R_{S2} + R_N + \dfrac{R_V Z_{IN}}{R_V + Z_{IN}}}$$

$$= \frac{1}{500\ \Omega + 25\ \text{K} + \dfrac{113\ \text{K} \times 1\ \text{M}}{113\ \text{K} + 1\ \text{M}}}$$

$$= \frac{-1}{127\ \text{K}} \cong -7.9\ \mu\text{A}$$

10. Calculate $C_1 = \dfrac{1}{2\pi F\left(R_{S1} + R_M + \dfrac{R_F Z_{\text{IN}}}{R_F + Z_{\text{IN}}}\right)}$

$$= \frac{1}{6.28(10)(127\ \text{K})} = 0.13\ \mu\text{F}$$

$$C_2 = \frac{1}{2\pi F\left(R_{S2} + R_N + \dfrac{R_V Z_{\text{IN}}}{R_V + Z_{\text{IN}}}\right)}$$

$$= \frac{1}{6.28(10)(127\ \text{K})} = 0.13\ \mu\text{F}$$

11. Calculate $I_{\text{OUT}} = V_{\text{OUT}}/R_L = 9$ vac/2 K = 4.5 mA.
12. Connect circuit and power supply to breadboard.
13. Adjust offset null.
14. Apply 1 vac peak in-phase signal at 1,000 Hz to inverting and noninverting inputs.
15. Monitor the output with an ac voltmeter. Results will be zero.
16. Apply 1 vac peak out-of-phase signal to inverting input and 1 vac peak in-phase signal to the noninverting input.
17. Monitor the output with an ac voltmeter. Results will be 9 vac in phase.
18. Apply 1 vac peak in-phase signal to inverting input and 1 vac peak out-of-phase signal to noninverting input.
19. Monitor the output with an ac voltmeter. Results will be 9 vac out of phase.

## SUMMING AMPLIFIER

Summing amplifiers, using operational amplifiers, are used for summing several signals to provide a single amplified output signal which may or may not be inverted depending on the type of design selected. These amplifiers are used in logic and control circuits and are of four basic types: dc summing inverting, ac summing inverting, dc summing noninverting, and ac summing noninverting. Before proceeding with the selection and design of any of these amplifiers, it is highly advisable to review the summing-point theory explained in the following paragraphs.

**Summing-Point Theory.** Ideal summing is accomplished when the sum point is a low impedance point. This simplifies the calculations and minimizes input interaction because the summed inputs do not load each other. The inverting summing amplifier is a good example of this because the inverting input (or sum point) is at ground potential because the noninverting input has been grounded.

The rationale is that negative feedback drives the inverting input in a direction that will reduce the differential between the two inputs. This keeps the inverting input at approximately ground potential for all inputs below the level required to saturate the amplifier output.

The noninverting summing amplifier is a good example of less-than-desirable summing, because the noninverting input must be at some finite resistance above ground. The result is more complex summing equations because of the interaction between inputs. The noninverting summer can be used to average input voltages.

As can be seen, the inverting summing amplifier is a better candidate for signal summing and should be favored unless phasing is important. Even if phasing is of concern, two inverting amplifiers in series may be a better solution than a single noninverting summing amplifier.

**DC Summing Inverting Amplifier Description (See figure 6-11)** As an introduction to the dc summing inverting amplifier, we must once again reach back to some op amp basics:

1. The summing points (terminals 2 and 3) are at virtual ground.
2. No current flows into the amplifier.
3. Feedback current is the same as the input current.
4. The difference voltage between terminals 2 and 3 is zero.

In the circuit shown in figure 6-11, there are three inputs, $V_{\text{IN1}}$, $V_{\text{IN2}}$, and $V_{\text{IN3}}$. These three inputs feed three resistors $R_M$, $R_N$, and $R_V$, which are connected to the summing point at the inverting terminal 2. The feedback resistor $R_F$ is also tied to the terminal 2.

There are three basic types of summers. They all have the same physical circuit setup. The first is the *summing inverter*. This circuit enlists equal input resistors $R_M$, $R_N$, and $R_V$. Indeed, the feedback resistance is also equal to these.

The second circuit is a *summing amplifier*. In this circuit the input

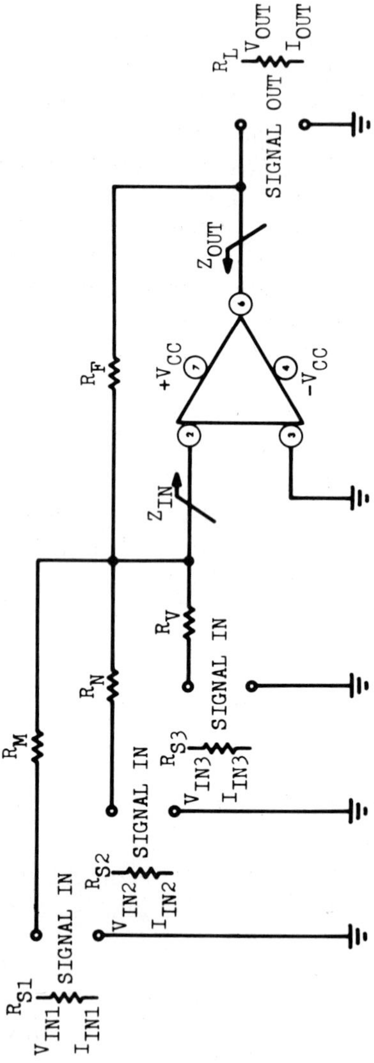


Figure 6-11 DC Summing Inverter

resistances are the same but the feedback resistance is selected to provide amplification. Resistor sizes can be different, which entails different gains for each input loop.

The third circuit is the *scaling adder*. In this circuit the input resistances are chosen for a scaled equal output for each input. Input currents, of course, would be equal.

Let's consider the circuit. From op amp theory, the following is equated:

$$I_{RM} + I_{RN} + I_{RV} = I_{RF}$$

Also

$$I_{RM} = \frac{V_{\text{IN1}}}{R_M}$$

$$I_{RN} = \frac{V_{\text{IN2}}}{R_N}$$

$$I_{RV} = \frac{V_{\text{IN3}}}{R_V}$$

$$I_{RF} = \frac{V_{\text{OUT}}}{R_F}$$

If the resistors are the same, the voltage gain is

$$A_v = -\frac{R_F}{R_M} = \text{unity}$$

The output voltage is the sum of the inputs:

$$V_{\text{OUT}} = V_{\text{IN1}} + V_{\text{IN2}} + V_{\text{IN3}}$$

If the resistors are different, the voltage output is still the sum except for different gain:

$$V_{\text{OUT}} = V_{\text{IN1}}\frac{R_F}{R_M} + V_{\text{IN2}}\frac{R_F}{R_N} + V_{\text{IN3}}\frac{R_F}{R_V}$$

The voltage gain can be calculated for each loop:

$$A_v = -\frac{R_F}{R_M}$$

or it may be summed:

$$A_v = \frac{-R_F\left[\dfrac{V_{\text{IN1}}}{R_M} + \dfrac{V_{\text{IN2}}}{R_N} + \dfrac{V_{\text{IN3}}}{R_V}\right]}{V_{\text{IN1}} + V_{\text{IN2}} + V_{\text{IN3}}}$$

Regardless of the type, the design uses a feedback resistor between output pin 6 and inverting pin 2.

Power is applied to pins 4 and 7. Resistors $R_M$, $R_N$, and $R_V$ provide voltage summing at the inverting input. The noninverting input pin 3 is grounded. This makes the effective input impedance zero at pin 2, since both inputs attempt to be equal.

Output impedance is typically 25 to 50 ohms and can be ignored. Input signals are normally less than 1.0 vdc, while output levels are from 0 to 12 vdc. Voltage gains are selective (typically under 50).

## DC SUMMING INVERTER—DESIGN PROCEDURES (See figure 6-11)

1. Select an operational amplifier that will provide the acceptable output.
2. Choose a dc power supply that will be compatible with the op amp.
3. Determine the source resistances $R_{S1}$, $R_{S2}$, and $R_{S3}$.
4. Choose $R_M$ to be large (rule of thumb—$R_M = 25\ R_{S1}$).
5. Choose $R_N$ to equal $R_M$.
6. Choose $R_V$ to equal $R_M$.
7. Consider that there is no voltage gain required:

$$A_v = \frac{R_F}{R_M} = \text{unity}$$

where $$R_F = R_M$$

**Note:** This design is for circuits that have unity gain. All resistances are the same and no amplification takes place.

8. Choose $R_F$ to equal $R_M$.
9. Calculate current in:

$$I_{\text{IN1}} = I_{\text{IN2}} = I_{\text{IN3}}$$

$$I_{\text{IN1}} = \frac{V_{\text{IN1}}}{R_{S1} + R_M}$$

10. Calculate current output:

$$I_{OUT} = \frac{V_{OUT}}{R_L}$$

11. Connect the circuit and the power supply on a breadboard.
12. Adjust offset null to zero the op amp if necessary.
13. Apply three positive dc signals to the inputs.
14. Monitor the output with a dc voltmeter.
15. Apply three negative dc signals to the inputs.
16. Monitor the output with a dc voltmeter.
17. Apply negative and positive dc signals to the three inputs.
18. Monitor the output with a dc voltmeter.

## DC SUMMING INVERTER—DESIGN EXAMPLE (See figure 6-11)

**Given:**

$$R_{S1} = 1\text{ K} \qquad V_{IN1} = \pm 1\text{ V}$$

$$R_{S2} = 1\text{ K} \qquad V_{IN2} = \pm 1\text{ V}$$

$$R_{S3} = 1\text{ K} \qquad V_{IN3} = \pm 1\text{ V}$$

$$R_L = 2\text{ K} \qquad V_{OUT\,MAX} = \pm 3\text{ V}$$

**Procedures:**

1. Choose $\mu$A741 op amp.
2. Choose $\pm 15$ vdc power supply.
3. Use 1 K for $R_{S1}$, $R_{S2}$, and $R_{S3}$.
4. Choose $R_M = 25\ R_{S1} = 25 \times 1\text{ K} = 25\text{ K}$.
5. Choose $R_N = R_M = 25$ K.
6. Choose $R_V = R_M = 25$ K.
7. Consider $A_v =$ unity.
8. Choose $R_F = R_M = 25$ K.
9. Calculate current in:

$$I_{IN1} = \frac{V_{IN1}}{R_M} = \frac{1\text{ V}}{25\text{ K}} = 40\ \mu\text{A}$$

$$I_{IN2} = 40\ \mu\text{A}$$

$$I_{IN3} = 40\ \mu\text{A}$$

10. Calculate current out:

$$I_{OUT} = \frac{V_{OUT}}{R_L} = \frac{3\text{ V}}{2\text{ K}} = 1.5\text{ mA}$$

11. Connect circuit and power supply to breadboard.
12. Adjust offset null.
13. Apply 1 V to $V_{IN1}$, $V_{IN2}$, and $V_{IN3}$.
14. Monitor −3 V output with dc voltmeter.
15. Apply −1 V to $V_{IN1}$, $V_{IN2}$, and $V_{IN3}$.
16. Monitor +3 V output with dc voltmeter.
17. Apply −1 V to $V_{IN1}$ and $V_{IN2}$. Apply +1 V to $V_{IN3}$.
18. Monitor +1 V output with dc voltmeter.

## DC SUMMING INVERTING AMPLIFIER—DESIGN PROCEDURES (See figure 6-11)

**Note:** The circuit for the dc summing amplifier is the same as the dc summing inverter. The difference in the feedback resistance value provides gain for each loop.

1. Select an operational amplifier that will provide the acceptable output.
2. Choose a dc power supply that will be compatible with the op amp.
3. Determine the source resistances $R_{S1}$, $R_{S2}$, and $R_{S3}$.
4. Choose $R_M$ to be large (rule of thumb—$R_M = 25\ R_{S1}$).
5. Choose $R_N$ to equal $R_M$.
6. Choose $R_V$ to equal $R_M$.
7. Calculate voltage gain:

$$A_{v1} = \frac{V_{OUT1}}{V_{IN1}} = -\frac{R_F}{R_M}$$

$$A_{v2} = \frac{V_{OUT2}}{V_{IN2}} = -\frac{R_F}{R_N}$$

$$A_{v3} = \frac{V_{OUT3}}{V_{IN3}} = -\frac{R_F}{R_V}$$

$$A_{v\,TOTAL} = \frac{V_{OUT}}{V_{IN1} + V_{IN2} + V_{IN3}}$$

8. Calculate feedback resistance $R_F$:

$$R_F = A_{v1} R_M \quad \text{(since gain in each loop is the same)}$$

9. Calculate current in:

$$I_{IN1} = \frac{V_{IN1}}{R_M}$$

$$I_{IN2} = \frac{V_{IN2}}{R_N}$$

$$I_{IN3} = \frac{V_{IN3}}{R_V}$$

$$I_{IN} = I_{IN1} + I_{IN2} + I_{IN3}$$

10. Calculate voltage output $V_{OUT}$:

$$V_{OUT} = A_{v1} V_{IN1} + A_{v2} V_{IN2} + A_{v3} V_{IN3}$$

11. Calculate current output:

$$I_{OUT} = \frac{V_{OUT}}{R_L}$$

12. Connect the circuit and the power supply on a breadboard.
13. Adjust offset null to zero the op amp if necessary.
14. Apply three positive dc signals to each input.
15. Monitor the output with a dc voltmeter.

## DC SUMMING INVERTING AMPLIFIER—DESIGN EXAMPLE (See figure 6-11)

**Given:**

$R_{S1} = 1$ K $\qquad V_{IN1} = 0.2$ vdc

$R_{S2} = 1$ K $\qquad V_{IN2} = 0.2$ vdc

$R_{S3} = 1$ K $\qquad V_{IN3} = 0.2$ vdc

$R_L = 2$ K $\qquad V_{OUT} = 6$ vdc

**Procedures:**

1. Choose $\mu$A741 op amp.
2. Choose $\pm 15$ vdc power supply.
3. Use 1 K for $R_{S1}$, $R_{S2}$, and $R_{S3}$.
4. Choose $R_M = 25\ R_{S1} = 25 \times 1\text{ K} = 25\text{ K}$.
5. Choose $R_N = R_M = 25\text{ K}$.
6. Choose $R_V = R_M = 25\text{ K}$.
7. Calculate voltage gain per loop:

$$A_v = \frac{V_{\text{OUT}}}{V_{\text{IN1}} + V_{\text{IN2}} + V_{\text{IN3}}}$$

$$A_v = \frac{6\text{ vdc}}{0.2\text{ vdc} + 0.2\text{ vdc} + 0.2\text{ vdc}}$$

$$A_v = \frac{6\text{ vdc}}{.6\text{ vdc}} = 10$$

8. Calculate feedback resistance:

$$R_F = -A_{v1} R_M \quad \left(\text{since } A_{v1} = -\frac{R_F}{R_M}\right)$$

$$R_F = -10 \times 25\text{ K} = 250\text{ K}$$

9. Calculate current in $I_{\text{IN}}$:

$$I_{\text{IN1}} = \frac{V_{\text{IN1}}}{R_M} = \frac{.2\text{ V}}{25\text{ K}} = 8\ \mu\text{A}$$

$$I_{\text{IN2}} = \frac{V_{\text{IN2}}}{R_N} = \frac{.2\text{ V}}{25\text{ K}} = 8\ \mu\text{A}$$

$$I_{\text{IN3}} = \frac{V_{\text{IN3}}}{R_V} = \frac{.2\text{ V}}{25\text{ K}} = 8\ \mu\text{A}$$

10. Calculate voltage output $V_{\text{OUT}}$ total:

$$V_{\text{OUT}} = A_{v1} V_{\text{IN1}} + A_{v2} V_{\text{IN2}} + A_{v3} V_{\text{IN3}}$$

$$V_{\text{OUT}} = 10(.2\text{ vdc}) + 10(.2\text{ vdc}) + 10(.2\text{ vdc})$$

$$V_{\text{OUT}} = 2\text{ vdc} + 2\text{ vdc} + 2\text{ vdc} = 6\text{ vdc}$$

11. Calculate current output:

$$I_{OUT} = \frac{V_{OUT}}{R_L} = \frac{6 \text{ vdc}}{2 \text{ K}} = 3 \text{ mA}$$

12. Connect circuit and power supply to breadboard.
13. Adjust offset null.
14. Apply 0.2 vdc to each input.
15. Measure 6 vdc at output with dc voltmeter.

## DC SCALING ADDER INVERTING AMPLIFIER—DESIGN PROCEDURES (See figure 6-11)

**Note:** The scaling adder inverting amplifier is designed so that the voltage gain of each input channel will provide the same voltage output. The sum of the voltage outputs will equal the total output. The sum of the currents through the input resistances will equal the current through the feedback resistance.

1. Select an operational amplifier that will provide the acceptable output.
2. Choose a dc power supply that will be compatible with the op amp.
3. Calculate the voltage gains on each input.

$$A_{v1} = \frac{\text{scale volts}}{V_{IN1}}$$

$$A_{v2} = \frac{\text{scale volts}}{V_{IN2}}$$

$$A_{v3} = \frac{\text{scale volts}}{V_{IN3}}$$

4. Calculate input resistances:

$$R_M = \frac{R_F}{A_{v1}}$$

$$R_N = \frac{R_F}{A_{v2}}$$

$$R_V = \frac{R_F}{A_{v3}}$$

5. Calculate input currents:

$$I_{RM} = \frac{V_{\text{IN1}}}{R_M}$$

$$I_{RN} = \frac{V_{\text{IN2}}}{R_N}$$

$$I_{RV} = \frac{V_{\text{IN3}}}{R_V}$$

6. Calculate feedback current:

$$I_{RF} = \frac{V_{\text{OUT}}}{R_F}$$

7. Ensure that feedback current $I_{RF}$ is equal to the sum of the input currents $I_{RM}$, $I_{RN}$, and $I_{RV}$.
8. Connect the circuit and the power supply on a breadboard.
9. Adjust offset null to zero op amp, if necessary.
10. Apply predetermined signals to inputs.
11. Monitor the output with a dc voltmeter.

## DC SCALING ADDER INVERTING AMPLIFIER—DESIGN EXAMPLE (See figure 6-11)

**Given:**

$$R_F = 60 \text{ K}$$

$$V_{\text{IN1}} = .1 \text{ vdc}$$

$$V_{\text{IN2}} = .2 \text{ vdc}$$

$$V_{\text{IN3}} = .3 \text{ vdc}$$

$$V_{\text{OUT}} = 6 \text{ vdc}$$

**Note:** Scale gain so that each input supplies 2 vdc of the output.

**Procedures:**

1. Choose $\mu$A741 op amp.
2. Choose $\pm 15$ vdc power supply.

3. Calculate:

$$A_{v1} = \frac{2\ V(\text{scale})}{V_{\text{IN1}}} = \frac{2\ \text{V}}{.1\ \text{V}} = 20$$

$$A_{v2} = \frac{2\ \text{V(scale)}}{V_{\text{IN2}}} = \frac{2\ \text{V}}{.2\ \text{V}} = 10$$

$$A_{v3} = \frac{2\ \text{V(scale)}}{V_{\text{IN3}}} = \frac{2\ \text{V}}{.3} = 6.67$$

4. Calculate input resistances:

$$R_M = \frac{R_F}{A_{v1}} = \frac{60\ \text{K}}{20} = 3\ \text{K}$$

$$R_N = \frac{R_F}{A_{v2}} = \frac{60\ \text{K}}{10} = 6\ \text{K}$$

$$R_V = \frac{R_F}{A_{v3}} = \frac{60\ \text{K}}{6.67} \cong 9\ \text{K}$$

5. Calculate input currents:

$$I_{RM} = \frac{V_{\text{IN1}}}{R_M} = \frac{.1\ \text{V}}{3\ \text{K}} = 33.3\ \mu\text{A}$$

$$I_{RN} = \frac{V_{\text{IN2}}}{R_N} = \frac{.2\ \text{V}}{6\ \text{K}} = 33.3\ \mu\text{A}$$

$$I_{RV} = \frac{V_{\text{IN3}}}{R_V} = \frac{.3\ \text{V}}{9\ \text{K}} = 33.3\ \mu\text{A}$$

6. Calculate feedback current:

$$I_{RF} = \frac{V_{\text{OUT}}}{R_F} = \frac{6\ \text{V}}{60\ \text{K}} = .1\ \text{mA}$$

7. Ensure that $I_{RF}$ equals the sum of the input currents $I_{RM}$, $I_{RN}$, and $I_{RV}$. $I_{RF} = 33.3\ \mu\text{A} + 33.3\ \mu\text{A} + 33.3\ \mu\text{A} = .1\ \text{mA}$
8. Connect circuit and power supply to breadboard.
9. Adjust offset null.
10. Apply .1 vdc to $V_{\text{IN1}}$, .2 vdc to $V_{\text{IN2}}$, and .3 vdc to $V_{\text{IN3}}$.
11. Measure 6 vdc at output with voltmeter.

**AC Summing Inverting Amplifier Description** (See figure 6-12) Components for the ac summing inverting amplifier are the same as for

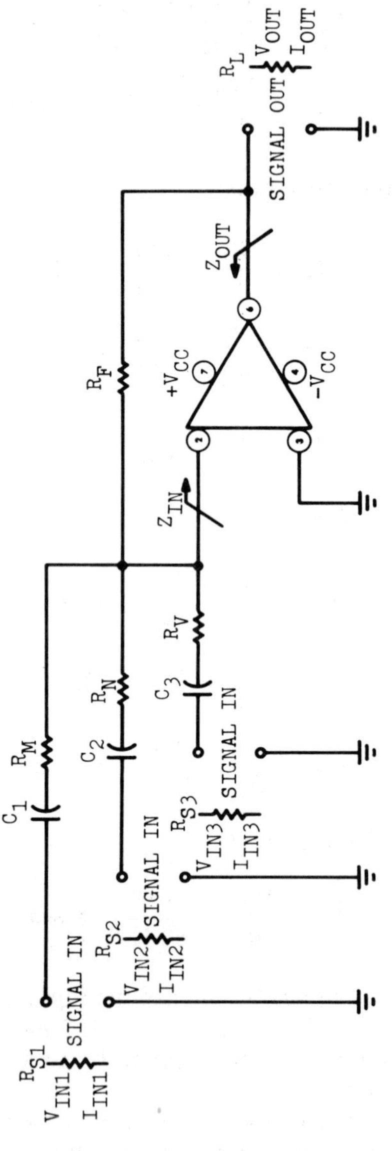


Figure 6-12 AC Summing Inverting Amplifier

the dc summing inverting amplifier except for the added ac coupling capacitors. Calculations for component size are similar to the dc summing inverting amplifier.

The design of this amplifier is rather simple, with a feedback resistor $R_F$ and resistors $R_M$, $R_N$, and $R_V$ provide a voltage divider for the inverting input. The noninverting input pin 3 is grounded. This makes the input impedance zero, since both inputs attempt to be equal. Output impedance is typically 25 to 50 ohms and can be ignored.

The inputs are coupled through capacitors $C_1$, $C_2$, and $C_3$. These capacitors are chosen by the frequency they may encounter. Input signals are between 0 and 0.5 vac peak while output levels are 0 to 10 vac peak. Voltage gains are low (typically under 50). Current gains are small (typically less than 500).

## AC SUMMING INVERTING AMPLIFIER—DESIGN PROCEDURES (See figure 6-12)

1. Select an operational amplifier that will provide the acceptable output.
2. Choose a dc power supply that will be compatible with the op amp.
3. Determine the source resistances $R_{S1}$, $R_{S2}$, and $R_{S3}$.
4. Choose $R_M$ to be large (rule of thumb—$R_M = 25\ R_{S1}$).
5. Choose $R_N$ to equal $R_M$.
6. Choose $R_V$ to equal $R_M$.
7. Calculate voltage gain per loop:

$$A_{v1} = \frac{V_{\text{OUT1}}}{V_{\text{IN1}}} = -\frac{R_F}{R_M}$$

$$A_{v2} = \frac{V_{\text{OUT2}}}{V_{\text{IN2}}} = -\frac{R_F}{R_N}$$

$$A_{v3} = \frac{V_{\text{OUT3}}}{V_{\text{IN3}}} = -\frac{R_F}{R_V}$$

$$A_v \text{ Total} = \frac{V_{\text{OUT}}}{V_{\text{IN1}} + V_{\text{IN2}} + V_{\text{IN3}}}$$

8. Calculate feedback resistance:

$$R_F = A_{v1}\, R_M \quad \text{(since gain in each loop is the same)}$$

9. Calculate current in:

$$I_{IN1} = \frac{V_{IN1}}{R_M}$$

$$I_{IN2} = \frac{V_{IN2}}{R_N}$$

$$I_{IN3} = \frac{V_{IN3}}{R_V}$$

10. Calculate voltage out:

$$V_{OUT} = A_v V_{IN1} + A_v V_{IN2} + A_v V_{IN3}$$

11. Calculate current output:

$$I_{OUT} = \frac{V_{OUT}}{R_L}$$

12. Calculate input coupling capacitors at a break frequency of 10 hertz:

$$C_1 = C_2 = C_3 = \frac{1}{2\pi F(R_{S1} + R_M)}$$

13. Connect the circuit and the power supply on a breadboard.
14. Adjust offset null to zero the op amp, if necessary.
15. Apply the same positive in-phase signal at the same frequency to the amplifier inputs.
16. Monitor the output with an ac voltmeter.

## AC SUMMING INVERTING AMPLIFIER—DESIGN EXAMPLE (See figure 6-12)

**Given:**

| | | |
|---|---|---|
| $R_{S1} = 1$ K | $V_{IN1} = .1$ vac peak | $f_L = 10\ H_z$ |
| $R_{S2} = 1$ K | $V_{IN2} = .1$ vac peak | |
| $R_{S3} = 1$ K | $V_{IN3} = .1$ vac peak | |
| $R_L = 2$ K | $V_{OUT} = 9$ vac | |

**Procedures:**

1. Choose $\mu$A741 op amp.
2. Choose $\pm 15$ vdc power supply.
3. Use 1 K for $R_{S1}$, $R_{S2}$, and $R_{S3}$.
4. Choose $R_M = 25\ R_{S1} = 25 \times 1\text{ K} = 25\text{ K}$.
5. Choose $R_N = R_M = 25\text{ K}$.
6. Choose $R_V = R_M = 25\text{ K}$.
7. Calculate voltage gain:

$$A_v = \frac{V_{\text{OUT}}}{V_{\text{IN1}} + V_{\text{IN2}} + V_{\text{IN3}}}$$

$$A_v = \frac{9\text{ vac}}{.1\text{ mvac} + .1\text{ mvac} + .1\text{ mvac}}$$

$$A_v = \frac{9\text{ vac}}{.3\text{ mvac}} = 30$$

8. Calculate feedback resistance:

$$R_F = A_{v1}\,R_M$$

$$R_F = 30 \times 25\text{ K}$$

$$R_F = 750\text{ K}$$

9. Calculate current in:

$$I_{\text{IN1}} = \frac{V_{\text{IN1}}}{R_M} = \frac{.1\text{ vac}}{25\text{ K}} = 4\ \mu\text{A}$$

$$I_{\text{IN2}} = \frac{V_{\text{IN2}}}{R_N} = \frac{.1\text{ vac}}{25\text{ K}} = 4\ \mu\text{A}$$

$$I_{\text{IN3}} = \frac{V_{\text{IN3}}}{R_V} = \frac{.1\text{ vac}}{25\text{ K}} = 4\ \mu\text{A}$$

10. Calculate voltage out total:

$$V_{\text{OUT}} = A_v\,V_{\text{IN1}} + A_v\,V_{\text{IN2}} + A_v\,V_{\text{IN3}}$$

$$V_{\text{OUT}} = 30(.1\text{ vac}) + 30(.1\text{ vac}) + 30(.1\text{ vac})$$

$$V_{\text{OUT}} = 3\text{ vac} + 3\text{ vac} + 3\text{ vac}$$

$$V_{\text{OUT}} = 9\text{ vac}$$

11. Calculate current output:

$$I_{OUT} = \frac{V_{OUT}}{R_L}$$

$$I_{OUT} = \frac{9 \text{ vac}}{2 \text{ K}}$$

$$I_{OUT} = 4.5 \text{ mA}$$

12. Calculate input coupling capacitors:

$$C_1 = C_2 = C_3 = \frac{1}{2\pi F(R_{S1} + R_M)}$$

$$C_1 = \frac{1}{6.28(10)(1 \text{ K} + 25 \text{ K})}$$

$$C_1 = 0.61\ \mu F$$

13. Connect circuit and power supply to breadboard.
14. Adjust offset null.
15. Apply .1 vac peak, 400 Hz signal at the same phase to the three inputs.
16. Monitor 9 vac out of phase at output with an ac voltmeter.

**DC Summing Noninverting Amplifier Description (See figure 6-13)** The reader may wish to review the description portion entitled "DC Summing Inverting Amplifier" prior to reading this section.

In the circuit shown in figure 6-13, there are two inputs to the circuit, $V_{IN1}$ and $V_{IN2}$. These inputs feed the two resistors $R_N$ and $R_V$, respectively. They are connected to the noninverting summing point. $R_G$ is a gain setting resistance. $R_M$ is a feedback divider resistor. $R_F$ is the feedback resistor.

The resistances on the front end of the amplifier are chosen to be equal. The feedback resistance $R_F$ is calculated from the gain formula. Voltage gain is:

$$A_v = \frac{V_{OUT}}{V_{INA}}$$

## DC SUMMING NONINVERTING AMPLIFIER— DESIGN PROCEDURES (See figure 6-13)

1. Select an operational amplifier that will provide the acceptable output.
2. Choose a dc power supply that will be compatible with the op amp.
3. Determine the source resistances $R_{S1}$ and $R_{S2}$.
4. Choose $R_M$ to be large (rule of thumb—$R_M = 10\ R_{S1}$).
5. Choose $R_N$ to equal $R_M$.
6. Choose $R_V$ to equal $R_M$.
7. Choose $R_G$ to equal $R_M$.
8. Calculate $V_{\text{INA}}$.

$$V_{\text{INA}} = \frac{V_{\text{IN1}}\left[\dfrac{R_G(R_V + R_{S2})}{R_G + (R_V + R_{S2})}\right]}{\left[\dfrac{R_G(R_V + R_{S2})}{R_G + (R_V + R_{S2})}\right] + R_N} + \frac{V_{\text{IN2}}\left[\dfrac{R_G(R_N + R_{S1})}{R_G + (R_N + R_{S1})}\right]}{\left[\dfrac{R_G(R_N + R_{S1})}{R_G + (R_N + R_{S1})}\right] + R_V}$$

9. Calculate voltage gain:

$$A_v = \frac{V_{\text{OUT}}}{V_{\text{INA}}}$$

10. Calculate feedback resistance.

$$R_F = (A_v - 1)(R_M)$$

from

$$A_v = \frac{R_M + R_F}{R_F}$$

11. Connect the circuit and the power supply on a breadboard.
12. Adjust offset null to zero the op amp if necessary.
13. Apply a positive dc signal to each input.
14. Monitor the output with a dc voltmeter.

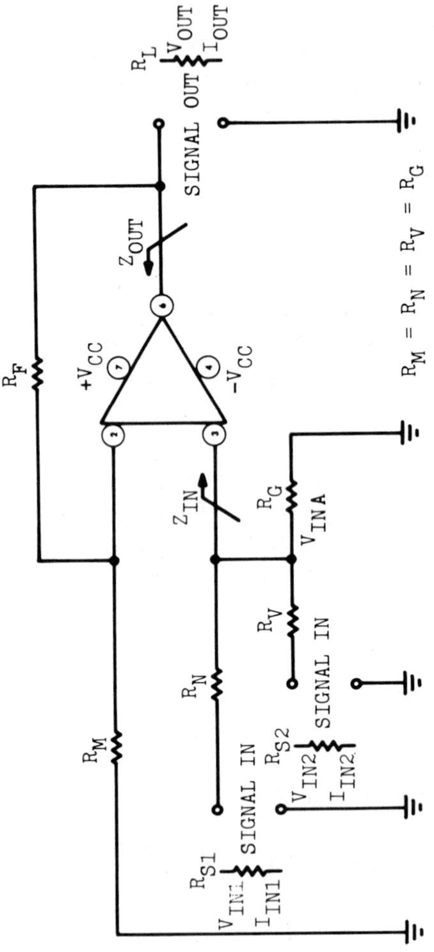


$$R_M = R_N = R_V = R_G$$

$$A_V = \frac{R_M + R_F}{R_M}$$

Figure 6-13 DC Summing Noninverting Amplifier

## DC SUMMING NONINVERTING AMPLIFIER—DESIGN EXAMPLE (See figure 6-13)

**Given:**

$$R_{S1} = 1\text{ K} \qquad V_{IN1} = 1.5\text{ vdc}$$

$$R_{S2} = 1\text{ K} \qquad V_{IN2} = 1.5\text{ vdc}$$

$$R_L = 2\text{ K} \qquad V_{OUT} = 10\text{ vdc}$$

**Procedures:**

1. Choose $\mu$A741 op amp.
2. Choose $\pm 15$ vdc power supply.
3. Use 1 K for $R_{S1}$ and $R_{S2}$.
4. Choose $R_M = 10\ R_{S1} = 10 \times 1\text{ K} = 10\text{ K}$.
5. Choose $R_N = R_M = 10\text{ K}$.
6. Choose $R_V = R_M = 10\text{ K}$.
7. Choose $R_G = R_M = 10\text{ K}$.
8. Calculate $V_{INA}$.

$$V_{INA} = \frac{1.5\left[\dfrac{10\text{ K}(10+1)\text{ K}}{10\text{ K}+(10+1)\text{ K}}\right]}{\left[\dfrac{10\text{ K}(10+1)\text{ K}}{10\text{ K}+(10+1)\text{ K}}\right]+10\text{ K}} + \frac{1.5\left[\dfrac{10\text{ K}(10+1)\text{ K}}{10\text{ K}+(10+1)\text{ K}}\right]}{\left[\dfrac{10\text{ K}(10+1)\text{ K}}{10\text{ K}+(10+1)\text{ K}}\right]+10\text{ K}} = 1.03\text{ vdc}$$

9. Calculate voltage gain.

$$A_v = \frac{V_{OUT}}{V_{INA}} = \frac{10}{1.03} \cong 9.71$$

10. Calculate feedback resistance.

$$R_F = (A_v - 1)(R_M)$$

$$R_F = (9.71 - 1)(10\text{ K})$$

$$R_F = 87.1\text{ K}$$

11. Connect circuit and power supply to breadboard.
12. Adjust offset null.
13. Apply 1.5 vdc to each input.
14. Measure 10 vdc at output with a dc voltmeter.

## DC DIRECT SUMMER (See figure 6-13)

The dc direct summer uses the same circuit as the dc summing noninverting amplifier. The dc direct summer does not amplify. It merely adds two or more voltages directly. The output voltage $V_{OUT}$ is equal to the sum of the input $V_{IN1}$ and $V_{IN2}$. The voltage gain ratio with this circuit is dependent on the number of inputs. It is significant that an equality exists between the input resistances and the feedback voltage divider.

$$\frac{R_F}{R_M} = \frac{R_G}{R_N} + \frac{R_G}{R_V}$$

and

$$R_G = R_N = R_V$$

This equality exists for any number of inputs. For example, the ratio $R_F/R_N$ must be 3 : 1 if there were three inputs, 4 : 1 for four inputs etc.

## DC DIRECT SUMMER—DESIGN PROCEDURES (See figure 6-13)

1. Select an operational amplifier that will provide the acceptable output.
2. Choose a dc power supply that will be compatible with the op amp.
3. Choose feedback resistance $R_F$ to be large, say 50 K to 100 K.
4. Determine number of inputs you wish to add.
5. Calculate metering resistance for the voltage divider $R_M = R_F/n$, where $n$ equals the number of inputs.
6. Choose $R_G$, $R_N$, $R_V$, to equal $R_F$.
7. Connect circuit and power supply on a breadboard.
8. Adjust offset null if necessary.
9. Apply dc voltage to inputs as selected.
10. Monitor output with a voltmeter.

## DC DIRECT SUMMER—DESIGN EXAMPLE (See figure 6-13)

**Given:**

$$V_{IN1} = 1.5 \text{ vdc}$$

$$V_{IN2} = 1.5 \text{ vdc}$$

$$V_{OUT} = V_{IN1} + V_{IN2}$$

$$R_F = \text{large, say } 50 \text{ K}$$

**Procedures:**

1. Choose $\mu$A741 op amp.
2. Choose $\pm 15$ vdc power supply.
3. Select $R_F = 50$ K.
4. Determine number of inputs as 2.
5. Calculate $R_M = \dfrac{R_F}{2} = \dfrac{50 \text{ K}}{2} = 25$ K.
6. Select $R_G = R_N = R_V = R_F = 50$ K.
7. Connect circuit and power supply on a breadboard.
8. Adjust offset null if necessary.
9. Apply 1.5 vdc to $V_{IN1}$ and $V_{IN2}$.
10. Monitor 3 vdc $V_{OUT}$ with a voltmeter.

**AC Summing Noninverting Amplifier Description (See figure 6-14)** The ac summing noninverting amplifier is identical to its dc counterpart, with the exception of the addition of ac coupling capacitors. This amplifier design uses a feedback resistor $R_F$ between output pin 6 and inverting input pin 2. Power is applied to pins 4 and 7. The $R_F$ resistor, along with the capacitor $C_3$, provides a voltage divider for feedback.

The first input $V_{IN1}$ is coupled across the resistor $R_N$ and capacitor $C_1$, and summed at resistor $R_G$. The second input signal $V_{IN2}$ is coupled through resistor $R_V$ and capacitor $C_2$ and also summed at resistor $R_G$. The two signals combine across $R_G$ as $V_{INA}$.

Voltage gain is small (typically under 50), while current gain is also small (typically under 100). The input capacitors are used primarily for coupling and are typically 6.8 $\mu$F for audio amplifiers. Input signals are typically below 10 vac peak.

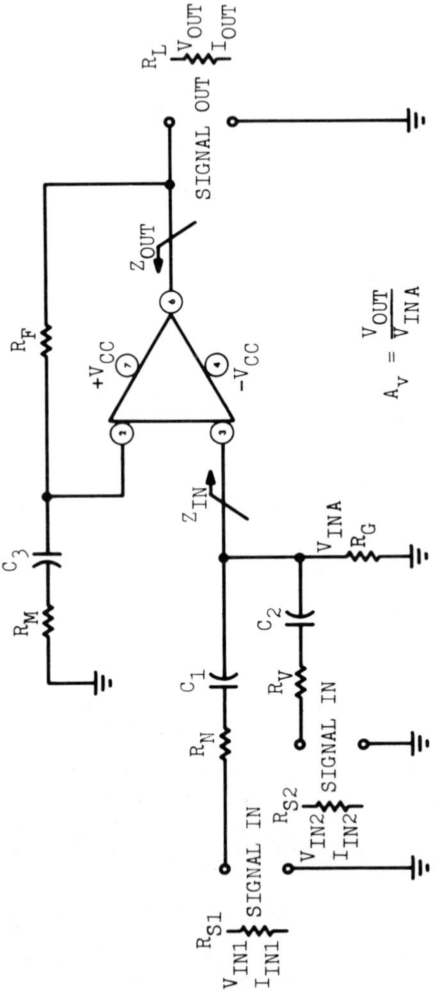


Figure 6-14 AC Summing Noninverting Amplifier

The ac summing noninverting amplifier is used in logic circuits and audio amplifiers. Voltage gain is:

$$A_v = \frac{V_{\text{OUT}}}{V_{\text{INA}}}$$

## AC SUMMING NONINVERTING AMPLIFIER—DESIGN PROCEDURES (See figure 6-14)

1. Select an operational amplifier that will provide the acceptable output.
2. Choose a dc power supply that will be compatible with the op amp.
3 Determine the source resistances $R_{S1}$ and $R_{S2}$.
4. Choose $R_M$ to be large (rule of thumb—$R_M = 10\ R_{S1}$).
5. Choose $R_N$ to equal $R_M$.
6. Choose $R_V$ to equal $R_M$.
7. Choose $R_G$ to be large ($100R_{S1}$).
8. Calculate $V_{\text{INA}}$:

$$V_{\text{INA}} = \frac{V_{\text{IN1}}\left[\dfrac{R_G(R_V + R_{S2})}{R_G + (R_V + R_{S2})}\right]}{\left[\dfrac{R_G(R_V + R_{S2})}{R_G + (R_V + R_{S2})}\right] + R_N} + \frac{V_{\text{IN2}}\left[\dfrac{R_G(R_N + R_{S1})}{R_G + (R_N + R_{S1})}\right]}{\left[\dfrac{R_G(R_N + R_{S1})}{R_G + (R_N + R_{S1})}\right] + R_V}$$

9. Calculate voltage gain:

$$A_v = \frac{V_{\text{OUT}}}{V_{\text{INA}}}$$

10. Calculate feedback resistance:

$$R_F = (A_v - 1)(R_M)$$

from

$$A_v = \frac{R_M + R_F}{R_F}$$

11. Choose coupling capacitors (typically 6.8 $u$F for audio frequencies).
12. Calculate capacitor $C_3$ for desire frequency break point (3 dB):

since
$$f_L = \frac{1}{2\pi R_M C_3}$$

then
$$C_3 = \frac{1}{2\pi f_L R_M}$$

13. Connect the circuit and the power supply on a breadboard.
14. Adjust offset null to zero the op amp if necessary.
15. Apply an ac signal of the same phase to each input.
16. Monitor the output with an ac voltmeter or oscilloscope.

## AC SUMMING NONINVERTING AMPLIFIER—DESIGN EXAMPLE (See figure 6–14)

**Given:**

$R_{S1} = 1$ K $\quad V_{\text{IN1}} = 1.5$ vac pk 400 Hz

$R_{S2} = 1$ K $\quad V_{\text{IN2}} = 1.5$ vac pk 400 Hz

$R_L = 2$ K $\quad V_{\text{OUT}} = 9$ vac pk 400 Hz

$f_L = 10$ Hz

**Procedures:**

1. Choose $\mu$A741 op amp.
2. Choose $\pm 15$ vdc power supply.
3. Use 1 K for $R_{S1}$ and $R_{S2}$.
4. Choose $R_M = 10\ R_{S1} = 10 \times 1\text{ K} = 10\text{ K}$.
5. Choose $R_N = R_M = 10$ K.
6. Choose $R_V = R_M = 10$ K.
7. Choose $R_G = 100\ R_{S1} = 100 \times 1\text{ K} = 100\text{ K}$.
8. Calculate $V_{\text{INA}}$:

$$V_{\text{INA}} = \frac{1.5\left[\dfrac{100\text{ K }(10+1)\text{ K}}{100\text{ K} + (10+1)\text{ K}}\right]}{\left[\dfrac{100\text{ K}(10+1)\text{ K}}{100\text{ K} + (10+1)\text{ K}}\right] + 10\text{ K}}$$

$$+\frac{1.5\left[\dfrac{100\text{ K}(10+1)\text{ K}}{100\text{ K}+(10+1)\text{ K}}\right]}{\left[\dfrac{100\text{ K}(10+1)\text{ K}}{100\text{ K}+(10+1)\text{ K}}\right]+10\text{ K}}$$

$$V_{\text{INA}}=\frac{14.86}{19.91}+\frac{14.86}{19.91}\cong 1.49\text{ vac}$$

9. Calculate voltage gain:

$$A_v=\frac{V_{\text{OUT}}}{V_{\text{INA}}}=\frac{9\text{ vac}}{1.49\text{ vac}}=6.04$$

10. Calculate feedback resistance:

$$R_F=(A_v-1)(R_M)$$

$$R_F=(6.04-1)(10\text{ K})$$

$$R_F=50.4\text{ K}$$

11. Choose coupling capacitors to be 6.8 $\mu$F.
12. Calculate feedback shunt capacitor:

$$C_3=\frac{1}{2\pi f_L R_M}=\frac{1}{2\times 3.14\times 10\times 10\text{ K}}\cong 1.6\ \mu\text{F}$$

13. Connect circuit and power supply to breadboard.
14. Adjust offset null.
15. Apply 1.5 vac peak 400 hertz signal of the same phase to each input.
16. Measure an output of 9 vac peak 400 hertz with an oscilloscope.

**Note:** This output is achieved whenever the sum of in-phase signals equals 3 vac.

## THE INTEGRATOR

The basic use of integrator circuits, using operational amplifiers, is in control systems. Two types that are commonly used are described in this section. These are the *dc summing integrator*, which integrates the sum of three input signals, and the *dc differential integrator*, which integrates the difference between two input signals. In both types, the rate of integration is a function of input voltage level and a selected time constant.

Calculations for these circuits cannot be simplified to elementary algebraic equations. Therefore, before going into the detail designs of these types of integrator circuits, readers should first review the next subsection to determine whether or not they are prepared to continue further in the study of integrator circuits.

**Integral Solution Example for Integrator Circuits** Calculations for the design of an integrator circuit require the solution of an integral calculus equation which cannot be simplified to an elementary form. This is necessary so that the component values can be determined for specific integration rates. Integration rate is a function of input voltage and the time constant of an $R_1C_1$ combination.

If voltage gain $A_v$ and input voltage $V_{IN}$ are known over this period of time, then $C_1$ can be determined by the following formula if $R_1$ is assigned a specific value. The $R_2C_2$ combination should then be made equal to the $R_1C_1$ combination for the differential case.

Example:

$$C_1 = \frac{\int (V_{IN2} = V_{IN1})\, dt}{A_v R_1 (V_{IN2} - V_{IN1})}$$

The solution of a similar integral of the same form is given for reference.

**Given:**

$$v = 1 \text{ volt (constant)}$$

$$t_0 = 0 \text{ seconds (lower limit)}$$

$$t = 10 \text{ seconds (upper limit)}$$

$$\int_{t_0}^{t} v\, dt = \frac{vt}{1}\Big]_{t_0}^{t} = vt\Big]_{t_0}^{t}$$

$$\int_{t_0}^{t} v\, dt = v[t - t_0]$$

$$\int_{0}^{10} v\, dt = vt\Big]_{0}^{10}$$

$$\int_{0}^{10} v\, dt = 1[10 - 0]$$

$$\int_{0}^{10} v\, dt = 10 \text{ volt seconds}$$

**Note:** The interval of the integral is from $t_0 \rightarrow t$, or 10 seconds. The operator term $dt$ means to take the antiderivative of the function in question. Having done so, the upper and lower limits are substituted into the variable $t$ and the lower limit is subtracted from the upper limit to get the value of the function.

DC Summing Integrator. (See figure 6–15) Design of the dc summing integrator, using an operational amplifier, requires moderate current gain (typically 1,000 to 3,000) and voltage gain up to near 100. Input signals to the summing integrator are fairly small (around 100 to 300 millivolts). Input impedances are very large (around 1 megohm). Output impedances are near zero (typically 25 to 50 ohms).

In the figure, the $C_1$ capacitor (nonpolarized and low-leakage type) provides the feedback from output pin 6 to inverting pin 2. The input summing resistors $R_1$, $R_2$, and $R_3$ provide signal scaling and sum with the capacitor $C_1$ at the inverting input. Power for the amplifier is applied to pins 4 and 7.

In this particular design, the output maximum is reached in 20 seconds. A switch $S_1$ is provided to short the input and output as a reset function.

## DC SUMMING INTEGRATOR—DESIGN PROCEDURES (See figure 6-15)

1. Select an operational amplifier that will provide the acceptable output.
2. Choose a dc power supply that will be compatible with the op amp.
3. Determine the source resistors $R_{S1}$, $R_{S2}$, and $R_{S3}$.
4. Calculate voltage gain required:

$$A_v = \frac{V_{OUT}}{V_{IN1} + V_{IN2} + V_{IN3}}$$

5. Choose $R_1 = R_2 = R_3$ (typically 100 K).
6. Calculate feedback capacitor $C_1$:

since $$A_v = \frac{\dfrac{-1}{R_1 C_1} \displaystyle\int_{t_0}^{t} (V_{IN1} + V_{IN2} + V_{IN3})\, dt}{V_{IN1} + V_{IN2} + V_{IN3}}$$

$$C_1 = \frac{-\displaystyle\int_{t_0}^{t} (V_{IN1} + V_{IN2} + V_{IN3})\, dt}{A_v R_1 (V_{IN2} + V_{IN2} + V_{IN3})}$$

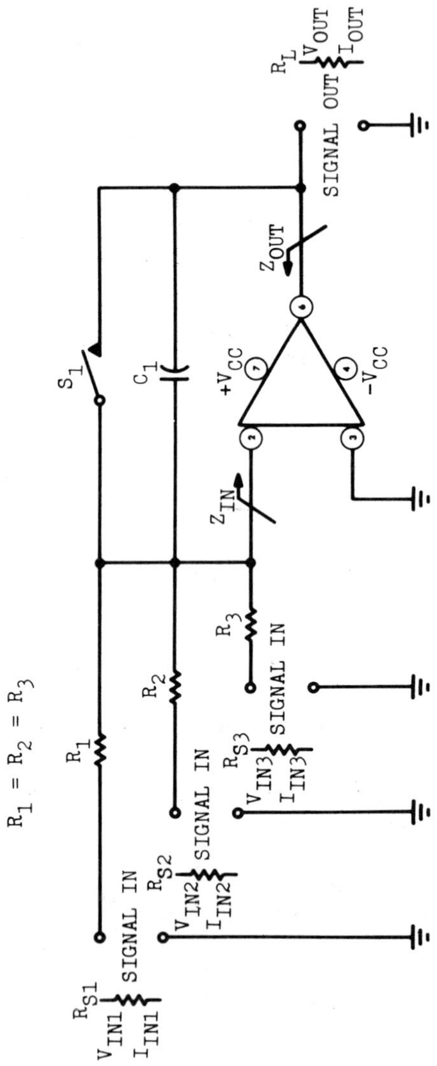


Figure 6-15 DC Summing Integrator

7. Choose switch $S_1$ (SPST).
8. Connect circuit and power supply to a breadboard.
9. Null $V_{IN}$ signals and adjust offset potentiometer for a null at the output with switch $S_1$ set to *on*.
10. Set switch $S_1$ to *off* and apply a low voltage simultaneously to $V_{IN1}$, $V_{IN2}$, and $V_{IN3}$. Monitor output on dc voltmeter in respect to time.
11. Set switch to *on* momentarily. Integrator output will drop to zero and then integrator will repeat its function.

## DC SUMMING INTEGRATOR—DESIGN EXAMPLE
(See figure 6-15)

**Given:**

| | |
|---|---|
| $V_{IN1} = 0.1$ vdc | $R_L = 2$ K |
| $V_{IN2} = 0.1$ vdc | $R_{S1} = 1$ K |
| $V_{IN3} = 0.1$ vdc | $R_{S2} = 1$ K |
| $V_{OUT} = 10$ vdc after 20 seconds | $R_{S3} = 1$ K |

**Procedures:**

1. Choose $\mu$A741 op amp.
2. Choose $\pm 15$ vdc power supply.
3. Use 1 K for $R_{S1}$, $R_{S2}$, and $R_{S3}$.
4. Calculate voltage gain required:

$$A_v = \frac{V_{OUT}}{V_{IN1} + V_{IN2} + V_{IN3}}$$

$$A_v = \frac{10 \text{ vdc}}{0.1 \text{ vdc} + 0.1 \text{ vdc} + 0.1 \text{ vdc}} \cong 33.4$$

5. Choose $R_1 = R_2 = R_3 = 100$ K.
6. Calculate feedback capacitor $C_1$:

$$C_1 = \frac{\int_0^{20} (0.1 + 0.1 + 0.1)\, dt}{(33.4)(100 \times 10^3)(0.1 + 0.1 + 0.1)} = \frac{0.3[20 - 0]}{1.002 \times 10^6}$$

$$= \frac{6}{1.002 \times 10^6} \cong 6\ \mu\text{F}$$

7. Choose switch $S_1$ (SPST).
8. Connect circuit and power supply to breadboard.
9. Null $V_{IN}$ signals and adjust offset potentiometer for a null at the output with switch $S_1$ set to *on*.
10. Set $S_1$ to *off* and apply 0.1 vdc simultaneously to $V_{IN1}$, $V_{IN2}$, and $V_{IN3}$. Monitor output of 10 vdc on dc voltmeter after an elapsed time of 20 seconds.
11. Set switch $S_1$ to on momentarily. Integrator output will drop to zero and integrator will repeat its function.

**Note:** Output will increase as a ramp function with a constant input. A ramp function is a signal output that increases linearly with time (see figure 6-16).

## DC DIFFERENTIAL INTEGRATOR (See figure 6-17)

Design of the differential dc integrator, using an operational amplifier, produces very high current gain (typically 25,000 to 75,000) and voltage gain of around 100. Input signals to the integrator are fairly small (around 100 to 300 millivolts). Input impedances are very large (around 1 megohm). Output impedances are near zero (typically 25 to 50 ohms).

In the figure, the $C_1$ capacitor (nonpolarized and low-leakage type) provides the feedback from output pin 6 to inverting pin 2. Resistor $R_1$ and capacitor $C_1$ provide a voltage divider for the inverting input. Resistor $R_2$ and capacitor $C_2$ (nonpolarized and low-leakage type) provide a voltage divider for the noninverting input. Power for the amplifier is applied to pins 4 and 7.

In this particular design, the output maximum is reached in 20 seconds.

## DC DIFFERENTIAL INTEGRATOR—DESIGN PROCEDURES (See figure 6-17)

1. Select an operational amplifier that will provide the acceptable output.
2. Choose a dc power supply that will be compatible with the op amp.
3. Determine the source resistances $R_{S1}$ and $R_{S2}$.
4. Calculate voltage gain required:

$$A_v = \frac{V_{OUT}}{V_{IN2} - V_{IN1}}$$

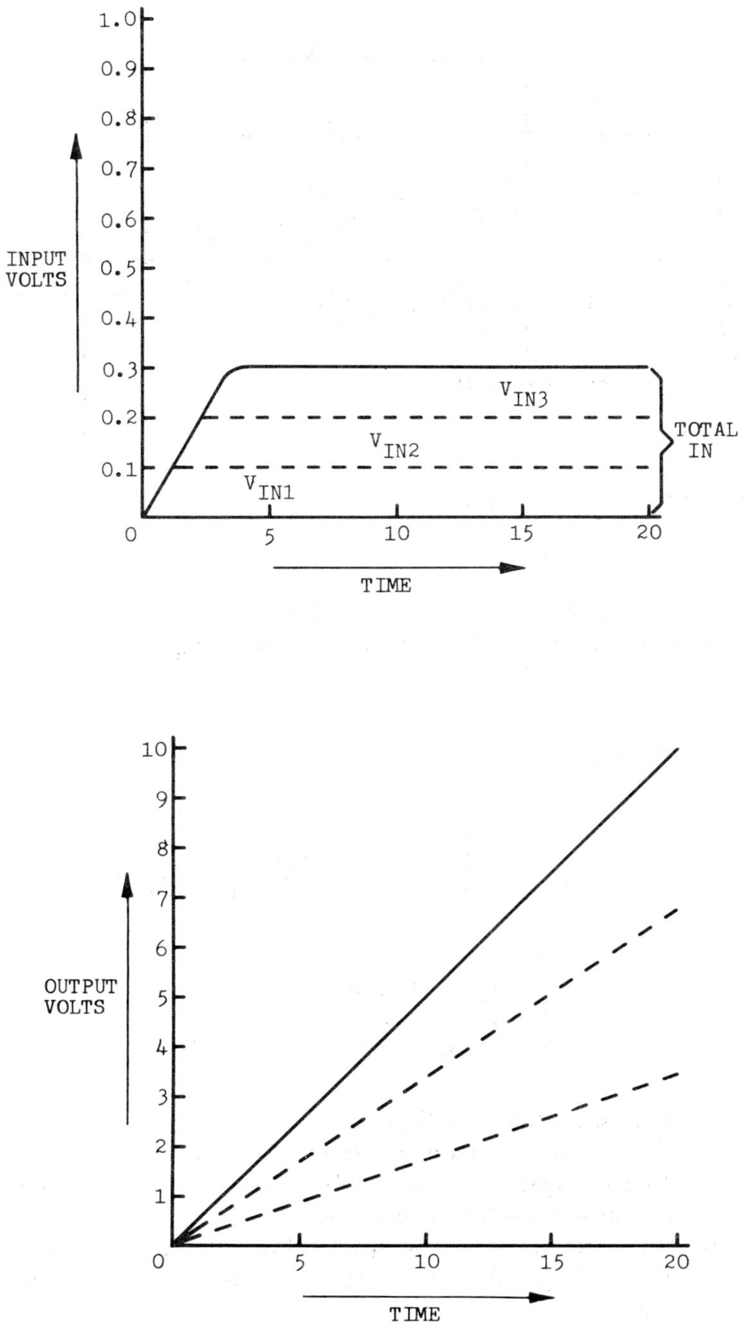


Figure 6-16 DC Summing Integrator Input and Output Graphs

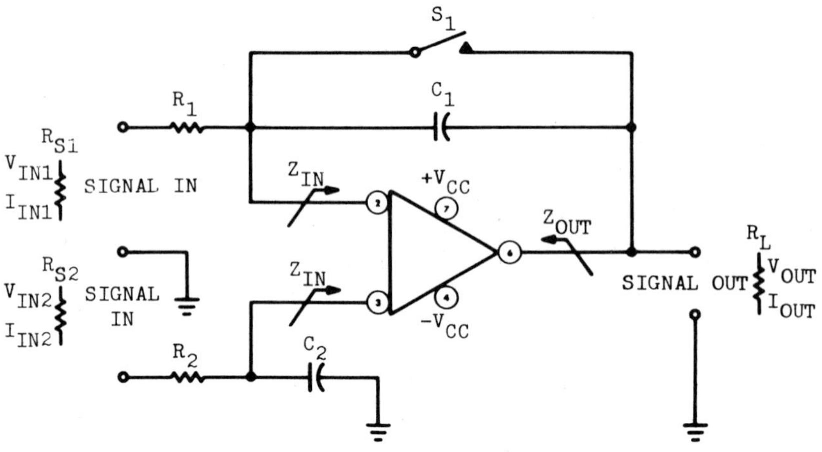


Figure 6-17 DC Differential Integrator

5. Choose $R_1 = R_2$ (typically 100 K).
6. Calculate feedback capacitor $C_1$ and voltage divider capacitor $C_2$:

Calculate

$$C_1 = C_2. \quad \left( \text{Since } A_v = \frac{\dfrac{1}{R_1 C_1} \displaystyle\int_{t_0}^{t} (V_{IN2} - V_{IN1})\, dt}{V_{IN2} - V_{IN1}} \right)$$

$$C_1 = \frac{\displaystyle\int_{0}^{20} (V_{IN2} - V_{IN1})\, dt}{A_v R_1 (V_{IN2} - V_{IN1})}$$

7. Choose switch $S_1$ (SPST).
8. Connect circuit and power supply to a breadboard.
9. Null $V_{IN}$ signals and adjust offset potentiometer for a null at the output with switch $S_1$ set to *on*.
10. Set switch $S_1$ to *off* and apply a low voltage simultaneously to $V_{IN1}$ and $V_{IN2}$. Monitor output with dc voltmeter in respect to time.
11. Set switch $S_1$ to *on* momentarily. Integrator output will drop to zero and then integrator will repeat its function.

## DC DIFFERENTIAL INTEGRATOR—DESIGN EXAMPLE (See figure 6-17)

**Given:**

| | |
|---|---|
| $V_{IN1} = 0.1$ vdc | $R_L = 2$ K |
| $V_{IN2} = 0.2$ vdc | $R_{S1} = 1$ K |
| $V_{OUT} = 10$ vdc after 20 seconds | $R_{S2} = 1$ K |

**Procedures:**

1. Choose $\mu$A741 op amp.
2. Choose $\pm 15$ vdc power supply.
3. Use 1 K for $R_{S1}$ and $R_{S2}$.
4. Calculate voltage gain required:

$$A_v = \frac{V_{OUT}}{V_{IN2} - V_{IN1}}$$

$$A_v = \frac{10 \text{ vdc}}{0.2 \text{ vdc} - 0.1 \text{ vdc}} = 100$$

5. Choose $R_1 = R_2 = 100$ K.
6. Calculate feedback capacitor $C_1$ and voltage divider capacitor $C_2$:

$$C_1 = C_2 = \frac{\int_0^{20} (0.2 - 0.1)\, dt}{(100)(100 \times 10^3)(0.2 - 0.1)} = \frac{0.1^{(20-0)}}{1 \times 10^6} = \frac{2}{1 \times 10^6}$$

$C_1$ & $C_2 = 2\ \mu$F

7. Choose switch $S_1$ (SPST).
8. Connect circuit and power supply to a breadboard.
9. Null $V_{IN}$ signals and adjust offset potentiometer for a null at the output with switch $S_1$ set to *on*.
10. Set $S_1$ to *off* and apply 0.1 vdc at $V_{IN1}$ and 0.2 vdc at $V_{IN2}$ simultaneously. Monitor output for 10 vdc on dc voltmeter after an elapsed time of 20 seconds.

**Note:** Output will increase as a ramp function, with a constant output. A ramp function is a signal output that increases linearly with time (see figure 6-18).

11. Set switch $S_1$ to *on* momentarily. Integrator output will drop to zero and the integrator will repeat its function.

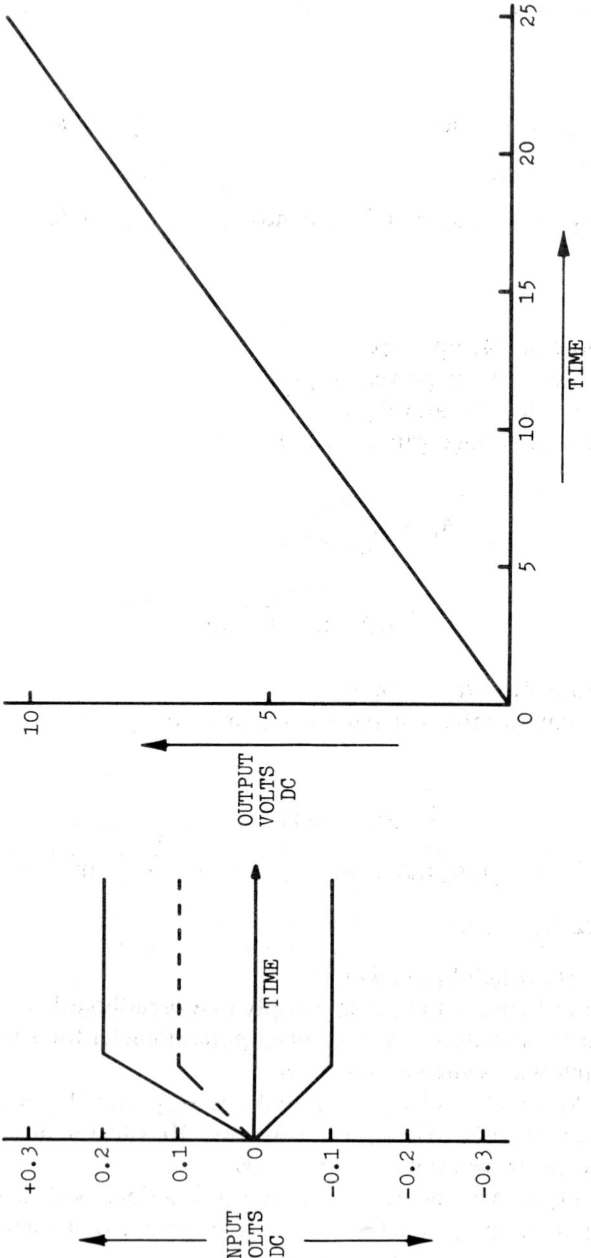


**Figure 6-18** DC Differentiator Input and Output Graphs

## THE DIFFERENTIATOR

The differentiator circuit is used in control circuits, and its function is mathematical. Its purpose is to produce an output that is the derivative of the input with respect to time, which is equivalent to the solution of a differential equation, using differentials. Use of differentials is necessary so that component values can be determined for a specific frequency-response range.

Calculations for the differentiator circuit cannot be simplified to elementary algebraic equations. Therefore, before going into the detail design of this circuit, readers should first review the next subsection to determine whether or not they are prepared to continue further in the study of the differentiator circuit.

## DIFFERENTIAL SOLUTION EXAMPLE FOR DIFFERENTIATOR CIRCUITS

The differentiator response is dependent on input impedance, feedback resistance, and input capacitance. The voltage gain $A_v$ is proportional to the input frequency. The approximate usable frequency range for a specific set of component values is 2 decades. A decade is an order of magnitude change such as $10 \rightarrow 100$ or $100 \rightarrow 1{,}000$. This bandwidth corresponds to a change in $A_v$ of approximately 100.

If voltage gain $A_v$ and input voltage $V_{IN}$ are known at a specific frequency, then the capacitor $C_1$ can be determined if $R_2$ is assigned a value, using the following formula:

$$C_1 = \frac{A_v \times V_{IN}}{-R_2 \dfrac{d(V_{IN})}{dt}}$$

**Note:** The minus sign in the denominator indicates that the output is 180° out of phase with the input and should be disregarded for calculations.

The differentiator output is linear as suggested earlier; however, capacitive effects cause the voltage gain to deteriorate at low frequency and at high frequency. The points where the voltage gain drops to 3 dB below the linear voltage gain curve (see figure 6-20) are considered the low- and high-frequency cutoff points. Therefore, the usable range of the differentiator in terms of frequency is between those two limits. The high- and low-frequency cutoff (3 dB) points can be calculated using the following formulas:

$$f_H = \frac{1}{2\pi(R_S + R_1)C_1}$$

$$f_L = \frac{1}{2\pi R_2 C_1}$$

The reader should be familiar with calculus and how to take the derivative of a function. The function may be any continuous function such as $y = x^2$ or $y = \sin x$. The derivative of each is presented below for reference:

Where the function is $y = x^2$, the derivative $\frac{dy}{dx} = 2x$

Where $$\frac{d}{dx}(x^2) = 2x^1 = 2x$$

Where the function is $y = \sin x$, the derivative $\frac{dy}{dx} = \cos x$

**Note:**

$$\frac{d}{dx}(\sin x) = \cos x \qquad \text{(only when } x \text{ is in radians)}$$

Where $$\left[\frac{d}{dx}(\sin x)\right]\left[\frac{d}{dx}(x)\right] = (\cos x)(1) = \cos x$$

Since control electronics frequently involve sinusoidal signals, the $y = \sin x$ function will be used to demonstrate the differentiator input and output. Mathematically, this is true if it can be shown that the function $y = \sin x$, when differentiated, is $\cos x$. If a measurement is performed we would expect the differentiator output phase to lead the differentiator input by 90 degrees. To provide insight, the solution of a similar differential equation is presented below:

**Given:**

$$V_{IN} = 2 \sin \omega t$$

then:

$$\frac{d}{dt}(2 \sin \omega t) = 2(\cos \omega t)\frac{d\omega t}{dt}$$

$$= 2(\cos 2\pi ft)\frac{d2\pi ft}{dt}$$

$$= 2(\cos 2\pi ft)(2\pi f)$$

$$= 4\pi f(\cos 2\pi ft)$$

This equation shows that the output is a cosine function and there is a phase shift of 90 degrees.

A means for determining the maximum amplitude of the cosine term at $t$ (time) = 0, or 0 angular degrees, is presented:

$$\begin{aligned}\frac{d}{dt}(2 \sin \omega t) &= 4\pi f \cos(2\pi ft)\\ &= 4\pi f \cos 2\pi f(0)\\ &= 4\pi f \cos(0)\\ &= 4\pi f(1.000)\\ &= 4\pi f \quad \text{(which represents the maximum voltage)}\end{aligned}$$

Since this is true, the output voltage will vary proportionately with frequency.

In the above equations, the operator $d/dt$ requires that the derivative be taken with respect to time. A table can be consulted for derivatives of other sinusoidal functions. The net effect of taking the derivative as shown above shifts the phase of the function by 90 degrees and increases the amplitude proportionately with frequency. Since we are interested in voltage amplitude for gain calculations, the cosine term can be evaluated at $t$ equal to zero, which is one (1).

The differentiator is exceptionally susceptible to noise, since gain increases with frequency, as can be seen from the formulas. Consequently, care must be taken to ensure accuracy in circuit calculations, and the best building practices should be exercised in construction.

**Differentiator Design (See figure 6-19)** Inputs to the differentiator are typically under 1 vdc and output voltages are typically under 10 vdc. Input impedances are near 1 megohm. Output impedances are 25 to 50 ohms. The resistor $R_2$ provides feedback between output pin 6 and inverting input pin 2. The capacitor $C_1$ and resistor $R_1$ provide a voltage divider and also couple the input signal into the amplifier. Power for the design is applied at pins 4 and 7. Noninverting pin 3 is grounded. Voltage gain varies with frequency.

## DIFFERENTIATOR—DESIGN PROCEDURES (See figure 6-19)

1. Select an operational amplifier that will provide the acceptable output.

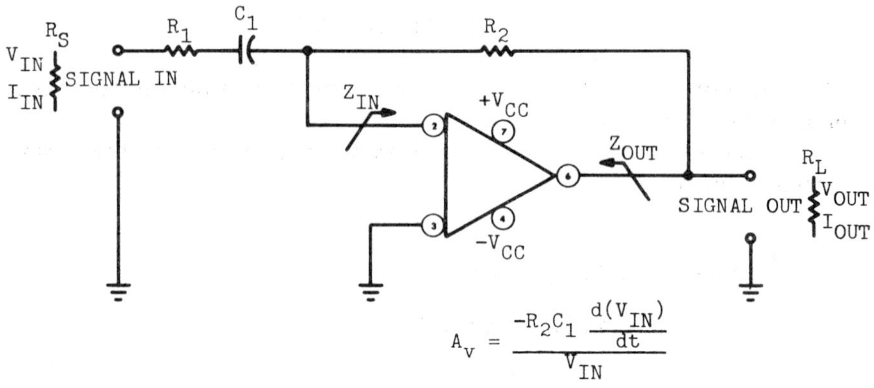


Figure 6-19 The Differentiator

2. Choose a dc power supply that will be compatible with the op amp.
3. Determine the source resistance $R_S$.
4. Calculate voltage gain required:

$$A_v = \frac{V_{\text{OUT}}}{V_{\text{IN}}}$$

5. Choose $R_1$ so that the sum of $R_S$ and $R_1$ equals 2 K.
6. Choose $R_2$ (typically 100 times $R_S + R_1$).
7. Calculate coupling capacitor $C_1$:

since $$A_v = \frac{-R_2 C_1 \dfrac{dV_{\text{IN}}}{dt}}{V_{\text{IN}}}$$

then $$C_1 = \frac{A_v V_{\text{IN}}}{-R_2 \dfrac{dV_{\text{IN}}}{dt}}$$

**Note:** Disregard minus signs in making calculations.

8. Calculate low break frequency of $C_1$:

$$f_L = \frac{1}{2\pi R_2 C_1}$$

9. Calculate high break frequency of $C_1$:

$$f_H = \frac{1}{2\pi(R_S + R_1)C_1}$$

10. Connect circuit and power supply on a breadboard.
11. Null $V_{IN}$ signal and adjust offset potentiometer for a null at the output.
12. Apply low voltage peak ac signal to inverting input.
13. Monitor dc output on oscilloscope in respect to time.

## DIFFERENTIATOR—DESIGN EXAMPLE (See figure 6-19)

**Given:**

$V_{IN} = 0.1(\sin \omega t)$ peak at 400 Hz $\qquad R_S = 1$ K

$V_{OUT} = 0.9(\cos \omega t)$ peak at 400 Hz $\qquad R_L = 2$ K

**Procedures:**

1. Choose $\mu$A741 op amp.
2. Choose $\pm 15$ vdc power supply.
3. Use 1 K for $R_S$.
4. Calculate voltage gain:

$$A_v = \frac{V_{OUT}}{V_{IN}}$$

$$A_v = \frac{0.9}{0.1} = 9$$

5. Choose $R_1 = 1$ K.
6. Calculate $R_2 = 100(R_S + R_1)$:

$$R_2 = 200 \text{ K}$$

7. Calculate coupling capacitor $C_1$:

$$C_1 = \frac{A_V V_{IN}}{-R_2 \dfrac{dV_{IN}}{dt}}$$

$$C_1 = \frac{0.9(\sin \omega t)}{200 \text{ K } (0.628)(400)(\cos \omega t)}$$

$$C_1 = \frac{0.9}{0.5024 \times 10^8} \cong 0.018\ \mu\text{F}$$

8. Calculate low break frequency of $C_1$:

$$f_L = \frac{1}{2\pi R_2 C_1}$$

$$f_L = \frac{1}{6.28(200\ \text{K})(0.018\ \mu\text{F})}$$

$$f_L \cong 44 \text{ hertz}$$

9. Calculate high break frequency of $C_1$:

$$f_H = \frac{1}{2\pi(R_S + R_1)C_1}$$

$$f_H = \frac{1}{6.28(2\ \text{K})(0.018\ \mu\text{F})}$$

$$f_H \cong 4423 \text{ hertz}$$

10. Connect circuit and power supply on a breadboard.
11. Null $V_{IN}$ signal and adjust offset potentiometer for a null at the output.
12. Apply 0.1 volt peak ac 400 Hz to input pin 2.
13. Monitor output of 0.9 volt peak ac 400 Hz with phase lagged 90 degrees with respect to the input, as observed on the oscilloscope.

**Note:** The differentiator is useful in control applications where the derivative of a function is required. See figure 6-20, showing output volts versus frequency in a differentiator.

## THE MULTIVIBRATOR

Multivibrator circuits, using operational amplifiers, are used in circuits requiring specific square-wave oscillations at specific or controlled frequencies. Basically, multivibrators are of three types: the *astable* (free-running) multivibrator, the *monostable* (one-shot), and the *steered bistable* (flip-flop) multivibrator design. These types are fairly stable, and exact frequencies can be obtained by the use of variable components.

The astable multivibrator has two equal outputs. The circuit output jumps from one to the other when power is turned on. The

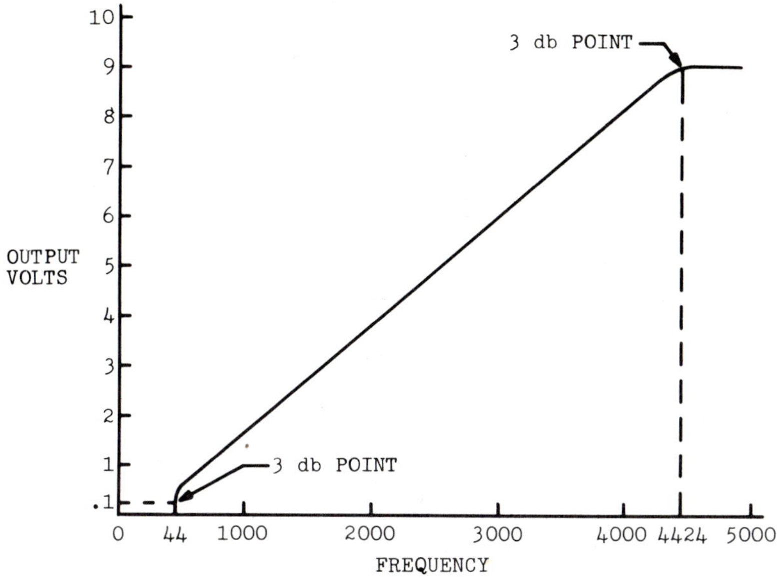


Figure 6-20 Output Volts Versus Frequency in a Differentiator

frequency of the jump is constant without any input signal. Hence the circuit is called astable or free-running.

The monostable multivibrator has a single stable state. When a signal is applied, the output changes for a predicted time, then returns to its original state. Therefore, the circuit is called a monostable or one-shot.

The bistable multivibrator has two stable state outputs. A signal of the proper polarity and amount will switch the polarity of the two output states. The circuit will then wait for the next signal. The circuit is two-state or bistable multivibrator (also called a flip-flop).

Astable (Free-Running) Multivibrator (See figure 6-21) The astable or free-running multivibrator is designed with either a steered input from some external sync circuit or can be designed with no input. Current calculations are not required. Voltage gain approaches infinity.

Power for the circuit is applied to pins 4 and 7. Feedback resistor $R_F$ is placed between output pin 6 and inverting input pin 2. Resistor $R_F$ and capacitor $C_1$ provide an $RC$ time constant for feedback and set the frequency. The resistors $R_F$ and $R_G$ are chosen equal to each other. $R_N$ resistance is chosen as a ratio to time constants and the voltage divider

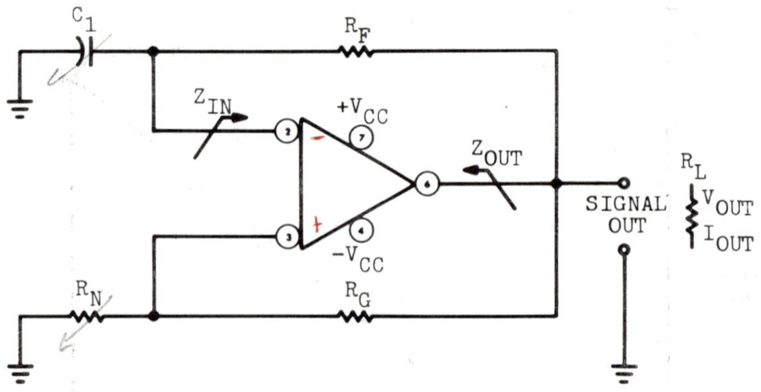


Figure 6-21 Astable Multivibrator

resistance $R_G$ (see design procedures). Input impedance is very high but not critical in this design. Output impedance is 25 to 50 ohms.

The circuit starts by itself and oscillates until power is removed. It can be made adjustable to ensure exact frequency by making the resistor $R_N$ or capacitor $C_1$ variable.

## ASTABLE MULTIVIBRATOR—DESIGN PROCEDURES (See figure 6-21)

1. Select a differential operational signal coupling amplifier with a high slew rate for square wave output.
2. Choose a dc power supply that will be compatible with the op amp.
3. Choose $R_F = R_G$ (typically 100 K).
4. Choose a free-running frequency.
5. Calculate timing capacitor $C_1$:

$$C_1 = \frac{1}{2\pi f\ R_F}$$

6. Calculate resistance $R_N$:

$$R_N = \frac{2\ \text{TC}\ (R_G)}{1\text{–}2\ \text{TC}}$$

where $\text{TC} = 63\%$

$$2\ \text{TC} = 86\%$$

7. Connect circuit and power supply on a breadboard.
8. Adjust offset null potentiometer to ensure that output is symmetrical.
9. Monitor output with an oscilloscope.

## ASTABLE MULTIVIBRATOR—DESIGN EXAMPLE
(See figure 6-21)

**Given:**

$$f_{set} = 200 \qquad R_L = 10\text{ K}$$

$$V_{OUT} = 13\text{ V peak square}$$

**Procedures:**

1. Choose $\mu$A741 op amp.
2. Choose $\pm 15$ vdc power supply.
3. Select $R_F = R_G = 100$ K.
4. Choose a set frequency:

$$f_{set} = 200$$

5. Calculate timing capacitor $C_1$:

$$C_1 = \frac{1}{2\pi f R_F}$$

$$C_1 = \frac{1}{6.28(200)(100\text{ K})}$$

$$C_1 \cong 0.008\ \mu\text{F}$$

6. Calculate resistance $R_N$:

$$R_N = \frac{2\text{ TC }(R_G)}{1\text{–}2\text{ TC}}$$

$$R_N = \frac{0.86(100\text{ K})}{0.14}$$

$$R_N \cong 614.3\text{ K}$$

7. Connect circuit and power supply on a breadboard.

**Note:** Multivibrator will oscillate when power is turned on.

8. Adjust offset null potentiometer if output is not symmetrical.
9. Monitor output for 13 volts square wave peak (saturation voltage) at 200 Hz, on an oscilloscope (see figure 6-22 for output waveshape).

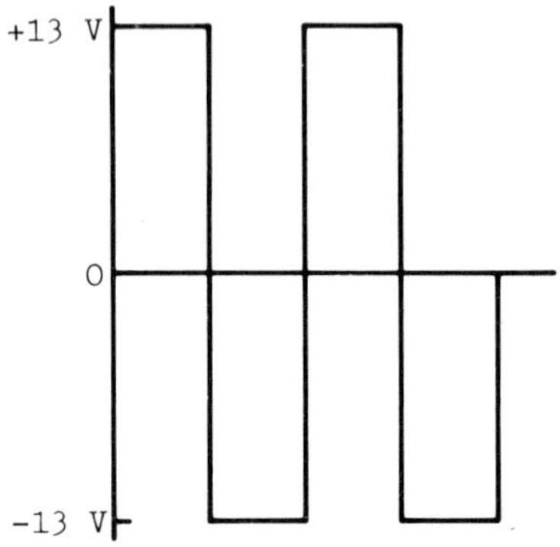


Figure 6-22 Output of Astable Multivibrator

**Steered Bistable (One-Shot) Multivibrator (See figure 6-23)** The bistable multivibrator is designed with either a steered input from some external sync circuit or with no input. The circuit we are using has a steered input. Current calculations are not required. Voltage gain is large.

Power for the circuit is applied to pins 4 and 7. Feedback is provided with a voltage divider resistor $R_3$ and a potentiometer $R_2$ placed

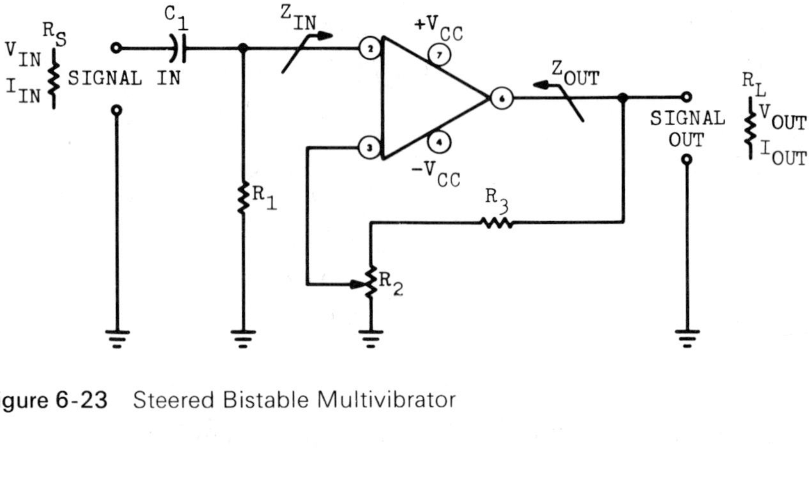


Figure 6-23 Steered Bistable Multivibrator

between output pin 6 and noninverting input pin 3. $R_3$ resistor is much larger than potentiometer $R_2$ (somewhere in the range of 1,000 times as large), while $R_2$ is chosen to be equal to the source resistance. Resistor $R_1$ is also equal to the source resistance. The resistor $R_1$ and capacitor $C_1$ provide a time constant for the input signal. The capacitor $C_1$ serves as a coupler. Input impedance is determined by $R_1$ resistor.

Output impedance is between 25 and 50 ohms. In this design the frequency is steered (controlled) by the input signal. With potentiometer $R_2$, the triggering level can be adjusted.

## STEERED BISTABLE MULTIVIBRATOR—DESIGN PROCEDURES (See figure 6-23)

1. Choose a differential op amp with a high slew rate for square wave output.
2. Choose a dc power supply that will be compatible with the op amp.
3. Select $R_S$ source resistor.
4. Choose $R_1$ equal to $R_S$.
5. Calculate coupling capacitor $C_1$ :

$$C_1 = \frac{1}{2\pi f(R_1 + R_S)}$$

6. Choose $R_2 = R_1$ (adjustable).
7. Choose $R_3 = 1{,}000\ R_2$.
8. Connect circuit and power supply on a breadboard.
9. Null $V_{IN}$ and adjust offset potentiometer for a null at the output.
10. Apply a signal at the input.
11. Monitor output with an oscilloscope.
12. Adjust $R_2$ as necessary to set trigger level of device.

## STEERED BISTABLE MULTIVIBRATOR—DESIGN EXAMPLE (See figure 6-23)

**Given:**

$f_{set} = 400$ hertz $\qquad R_L = 10$ K

$V_{OUT} = 13$ volts square wave $\qquad R_S = 500\ \Omega$

**Procedures:**

1. Choose $\mu$A741 op amp.
2. Choose $\pm 15$ vdc power supply.
3. Select $R_S = 500\ \Omega$.
4. Choose $R_1 = R_S = 500\ \Omega$.
5. Calculate coupling capacitor $C_1$:

$$C_1 = \frac{1}{2\pi f(R_1 + R_S)}$$

$$C_1 = \frac{1}{6.28(400)(500 + 500)}$$

$$C_1 = 0.398\ \mu F$$

6. Choose $R_2 = R_1 = 500\ \Omega$ (adjustable).
7. Choose $R_3 = 1{,}000\ R_2$
   $R_3 = 1{,}000 \times 500$
   $R_3 = 500$ K
8. Connect circuit and power supply on a breadboard.
9. Null $V_{IN}$ signal and adjust offset potentiometer for a null at the output.
10. Apply 1 vac 400 Hz at input.
11. Monitor output for 13 volts square wave peak (saturation voltage) at 400 Hz on an oscilloscope. Note waveshapes in figure 6-24.
12. Adjust $R_2$ as necessary to set trigger level of device.

Monostable (Flip-Flop) Multivibrator (See figure 6-25) The monostable multivibrator has only one stable output state. The circuit output stays in that state until a signal is applied. If the signal is the correct polarity and amount, the circuit output changes for a predicted period of time, then returns to its original state.

Let's consider the circuit in figure 6-25. The diode $CR_1$ clamps the inverting input of the amplifier at around a diode voltage drop (+.6 V). This causes the amplifier to sit at a latched positive state. When a negative $V_{IN}$ is coupled to the noninverting input that is greater than +.6 V, it causes the output to flip to the negative output state. The output is fed to the input through the feedback resistor $R_F$. The input cannot immediately follow because of the time constant $R_F C_1$. As the voltage across $C_1$ rises to the diode drop of $CR_1$, the output once again flips positive because of the feedback.

$C_2 R_3$ is a triggering network. $R_1 R_2$ sets the triggering sensitivity.

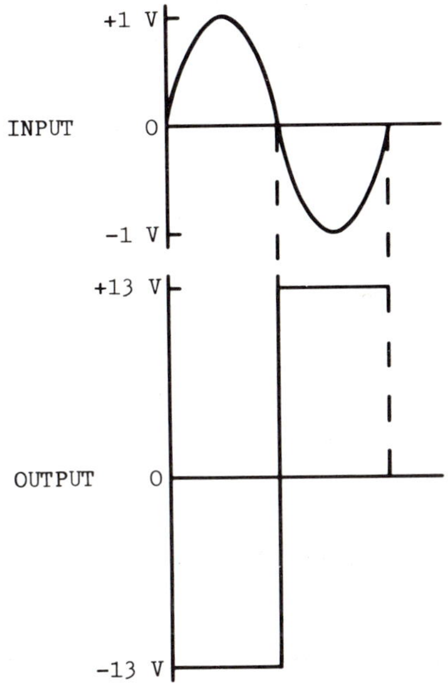


Figure 6-24 Waveshapes for Steered Bistable Multivibrator

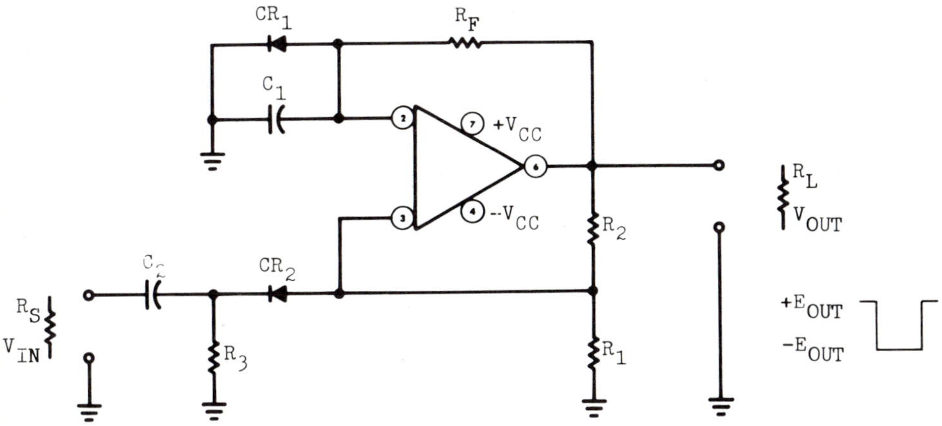


Figure 6-25 Monostable Multivibrator

The ratio sets the voltage level at the noninverting input. The time of the output pulse is determined by $R_F C_1$ if $V_{OUT}$ is greater that 0.6 V:

$$T = R_F C_1 \log_e\left(\frac{1}{1-\beta}\right)$$

where $\beta$ = triggering sensitivity

$$\beta = \frac{R_1}{R_1 + R_2}$$

If $\beta$ is made to be 0.632, then the log term of the last equation is equal to unity and the time $R_F C_1$ becomes the output pulse time:

$$T = R_F C_1$$

## MONOSTABLE MULTIVIBRATOR—DESIGN PROCEDURES (See figure 6-25)

1. Choose an op amp with a high slew rate for pulse wave output.
2. Choose a dc power supply that will be compatible with the op amp.
3. Select $R_S$ source resistor.
4. Determine the time ($T$) of the output pulse.
5. Select feedback resistance $R_F$ to be 100 K (typically).
6. Calculate the time constant capacitor $C_1$:

$$C_1 = \frac{T}{R_F}$$

7. Select capacitor $C_2$ = 10% $C_1$ (typically).
8. Select $R_3$ to be large (typically 500 K).
9. Select $R_1$ to be 10% $R_3$ (typically).
10. Determine trigger sensitivity ($\beta$) to be 0.632.
11. Calculate $R_2$:

Since
$$\beta = \frac{R_1}{R_1 + R_2}$$

$$R_2 = \frac{R_1(1-\beta)}{\beta}$$

12. Choose diodes $CR_1$ and $CR_2$ to be able to withstand feedback current flow and input current flow, respectively.

13. Connect circuit and power supply on a breadboard.
14. Adjust offset potentiometer for a null at the output.
15. Apply a negative input pulse below 0.6 V. Monitor output with an oscilloscope and time.

## MONOSTABLE MULTIVIBRATOR—DESIGN EXAMPLE (See figure 6-25)

**Given:**

pulse time = 2 msec $\quad R_L = 10$ K

$R_S = 1$ K

**Procedures:**

1. Choose $\mu$A741 op amp.
2. Choose $\pm 15$ vdc power supply.
3. Select $R_S = 1$ K.
4. Choose 2 msec pulse time.
5. Select $R_F = 100$ K.
6. Calculate $C_1$:

$$C_1 = \frac{T}{R_F}$$

$$C_1 = \frac{2\text{ msec}}{100\text{ K}}$$

$$C_1 = .02\ \mu\text{F}$$

7. Select $C_2 = 10\%\ C_1$
   $C_2 = .1 \times .02\ \mu$F
   $C_2 = .002\ \mu$F
8. Select $R_3 = 500$ K
9. Select $R_1 = 10\%\ R_3$
   $R_1 = .1 \times 500$ K
   $R_1 = 50$ K
10. Determine trigger sensitivity ($\beta$) to be 0.632.
11. Calculate $R_2$:

$$R_2 = \frac{R_1\,(1 - \beta)}{\beta}$$

$$R_2 = \frac{50\ \text{K}\ (1 - 0.632)}{0.632}$$

$$R_2 = \frac{50\ \text{K}\ (0.368)}{0.632} \cong 29\ \text{K}$$

12. Select $CR_1$ and $CR_2$ to be 1N914 diodes.
13. Connect circuit and power supply on a breadboard.
14. Adjust offset null.
15. Apply a negative input pulse below 0.6 V. Monitor output with an oscilloscope and time.

## THE COMPARATOR

The purpose of comparator circuits, using operational amplifiers, is to clamp large-output dc voltage levels with the use of smaller-control dc levels, while comparing the input with a reference voltage. These comparators, of five basic types which are explained and illustrated in this section, are used in logic or automatic control circuits for which specific clamping (vdc) levels are required. (The word "clamping" in electrical terms means to hold a voltage level at some specific value. This level can be positive, negative, or zero, such as +4 V, −3 V, or 0.)

Current and voltage gain calculations are not necessary for any of these amplifier designs, because they are used only as voltage comparators.

**Positive Clamped Simple Comparator** (See figure 6-26) Although current and voltage gain calculations are not necessary for this circuit design, voltage gain should be large so that the open-loop slew rate is acceptable.

Power for the positive clamped simple comparator is applied to pins 4 and 7. Noninverting pin 3 is grounded. A diode is placed in the

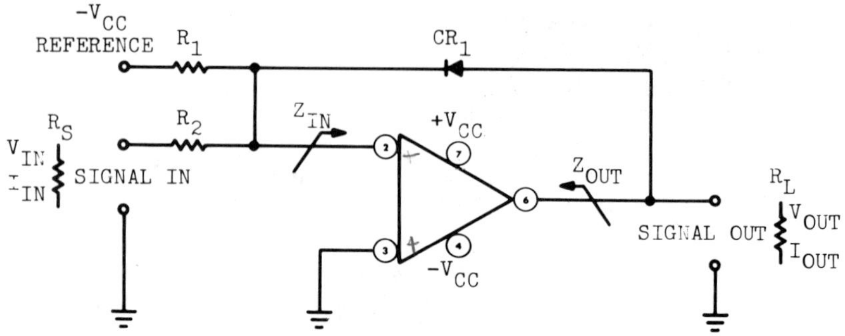


Figure 6-26 Positive Clamped Simple Comparator

feedback loop between output pin 6 and inverting pin 2. The diode is placed in such a manner that the circuit is open-loop for positive input levels and closed-loop for negative input levels. The reference $-V_{CC}$ voltage is compared through resistor $R_1$ with the variable $V_{IN}$ (input volts) through resistor $R_2$.

When $V_{IN}$ reaches desired clamping levels, the output changes state. A wide range of inputs may be used without protection devices for external circuits.

## POSITIVE CLAMPED SIMPLE COMPARATOR—DESIGN PROCEDURES (See figure 6-26)

1. Choose an operational amplifier that will provide the desirable characteristics of the circuit (high open-loop gain).
2. Choose a dc power supply that will be compatible with the op amp.
3. Choose $R_1$ (typically 100 K).
4. Choose $V_{IN}$ voltage (where $V_{IN}$ is input clamping level):
   therefore: $V_{OUT} = V_{CC} + 3$ volts (or saturation voltage) when $V_{IN}$ is greater than clamp level
   $V_{OUT} = 0.6$ volt when $V_{IN}$ is less than clamping level (Note that the 0.6 volt is the feedback diode voltage drop)
5. Calculate $R_2 = \dfrac{R_1 V_{IN}\ \text{(clamping level)}}{V_{CC}}$
6. Choose diode $CR_1$ to withstand feedback current (minor).
7. Connect circuit and power supply on a breadboard.
8. Remove $-V_{CC}$ reference and adjust offset null.
9. Apply $-V_{CC}$ reference and vdc above clamping level to the inverting input.
10. Monitor output with dc voltmeter.
11. Apply $V_{IN}$ at less than clamping level.
12. Monitor output with a dc voltmeter.

## POSITIVE CLAMPED SIMPLE COMPARATOR—DESIGN EXAMPLE (See figure 6-26)

**Given:**

$V_{IN} = 2$ vdc (clamping level) $\qquad R_L = 2$ K

$V_{OUT} = 0.6$ vdc (clamping level)

**Procedures:**

1. Choose $\mu$A741 op amp.
2. Choose $\pm 15$ vdc power supply.
3. Choose $R_1 = 100$ K (typical).
4. Choose $V_{IN} = 2$ vdc.
5. Calculate $R_2$:

$$R_2 = \frac{R_1\ V_{IN}\ (\text{clamping level})}{V_{CC}}$$

$$R_2 = \frac{100\ \text{K}\ (2\ \text{vdc})}{15} = 13.3\ \text{K}$$

6. Choose 1N4001 for $CR_1$.
7. Connect circuit and power supply on a breadboard.
8. Remove $-V_{CC}$ reference, null $V_{IN}$ signal, and adjust offset potentiometer for a null at the output.
9. Apply $-V_{CC}$ reference ($-15$) and $V_{IN}$ of 2.1 vdc to inverting input.
10. Monitor output of $-12$ vdc (saturation voltage) on voltmeter.
11. Apply $V_{IN}$ of 1.9 vdc.
12. Monitor output of 0.6 vdc on voltmeter.
    **Note:** See figure 6-27 for input versus output waveforms.

Negative Clamped—Simple Comparator (See figure 6-28) Although current and voltage gain calculations are not necessary for this circuit design, voltage gain should be large so that the open-loop slew rate is acceptable.

Power for the clamped comparator is applied to pins 4 and 7. Noninverting pin 3 is grounded. A diode is placed in the feedback loop between output pin 6 and inverting input pin 2. The diode is placed in such a manner that the circuit is open-loop for negative input levels and closed-loop for positive input levels. The reference $-V_{CC}$ voltage is compared through resistor $R_1$ with the variable $V_{IN}$ (input volts) through resistor $R_2$.

When $V_{IN}$ reaches desired clamping levels, the output changes state. A wide range of inputs may be used without protection devices for external circuits.

## NEGATIVE CLAMPED SIMPLE COMPARATOR—DESIGN PROCEDURES (See figure 6-28)

1. Choose an operational amplifier that will provide the desirable characteristics of the circuit (high open-loop gain).

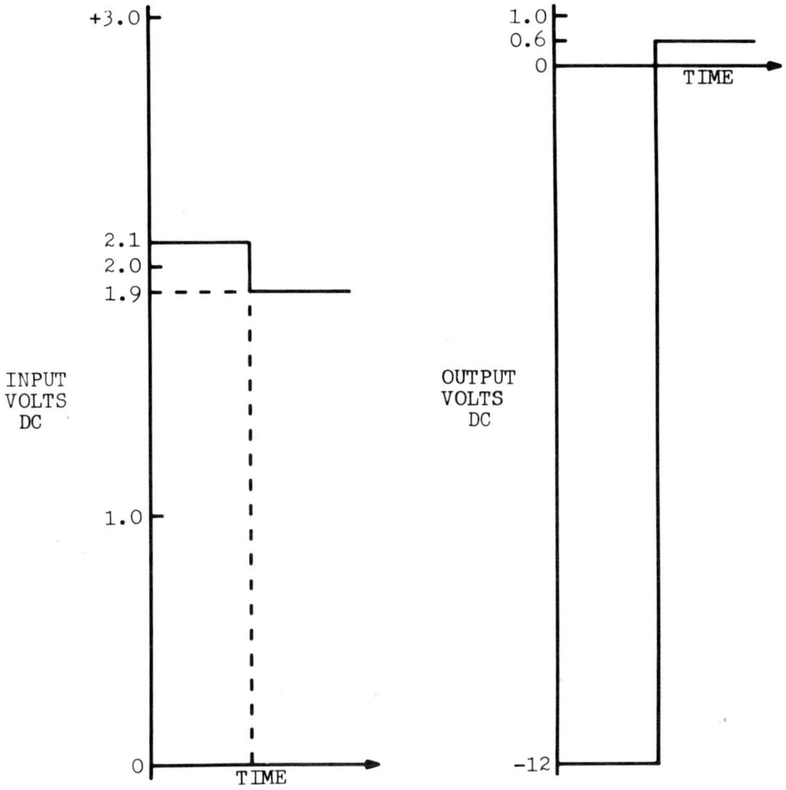


NOTE: WHEN INPUT IS CHANGED FROM 2.1 VDC TO 1.9 VDC, THE OUTPUT CLAMPS AT +0.6 VDC.

**Figure 6-27** Input Versus Output Waveforms, Positive Clamped Simple Comparator

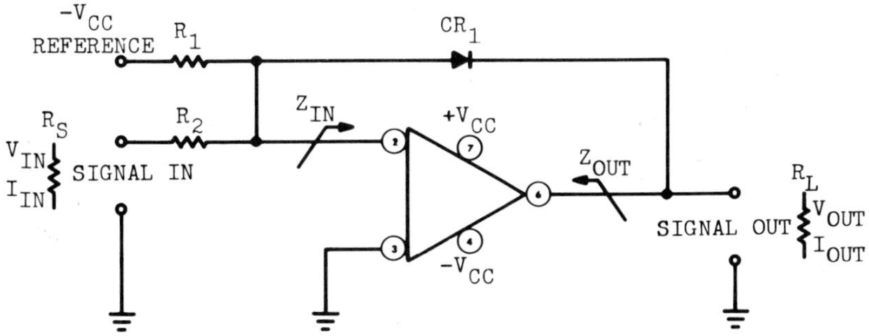


**Figure 6-28** Negative Clamped Simple Comparator

2. Choose a dc power supply that will be compatible with the op amp.
3. Choose $R_1$ (typically 100 K).
4. Choose $V_{IN}$ (where $V_{IN}$ is input clamping level):
   therefore: $V_{OUT} = V_{CC} - 3$ volts (or saturation voltage) when $V_{IN}$ is less than clamping level
   $V_{OUT} = 0.6$ volt when $V_{IN}$ is greater than clamping level (Note that the $-0.6$ volt is the feedback diode voltage drop)
5. Calculate $R_2 = \dfrac{R_1 V_{IN} \text{ (clamping level)}}{V_{CC}}$
6. Choose diode $CR_1$ to withstand feedback current (minor).
7. Connect circuit and power supply on a breadboard.
8. Remove $-V_{CC}$ reference and adjust offset null.
9. Apply $-V_{CC}$ reference and vdc above clamping level to the inverting input.
10. Monitor output with dc voltmeter.
11. Apply $V_{IN}$ at less than clamping level.
12. Monitor output with a dc voltmeter.

## NEGATIVE CLAMPED SIMPLE COMPARATOR—DESIGN EXAMPLE (See figure 6-28)

**Given:**

$V_{IN} = 1$ vdc (clamping level) $\qquad R_L = 2$ K

$V_{OUT} = -0.6$ vdc (clamping level)

**Procedures:**

1. Choose $\mu$A741 op amp.
2. Choose $\pm$vdc power supply.
3. Choose $R_1 = 100$ K (typical).
4. Choose $V_{IN} = 1$ vdc (clamping level).
5. Calculate $R_2$:

$$R_2 = \frac{R_1 V_{IN} \text{ (clamping level)}}{V_{CC}}$$

$$R_2 = \frac{100 \text{ K } (1 \text{ vdc})}{15} = 6.7 \text{ K}$$

6. Choose 1N4001 for $CR_1$.
7. Connect circuit and power supply on a breadboard.
8. Remove $-V_{CC}$ reference, null $V_{IN}$ signal, and adjust offset potentiometer for a null at the output.
9. Apply $-V_{CC}$ reference ($-15$ vdc) and $V_{IN}$ of 1.1 vdc to input.
10. Monitor output of $-0.6$ vdc on voltmeter.
11. Apply $V_{IN}$ of 0.9 vdc.
12. Monitor output of 12 vdc saturation voltage on voltmeter.

**Note:** See figure 6-29 for input versus output waveforms.

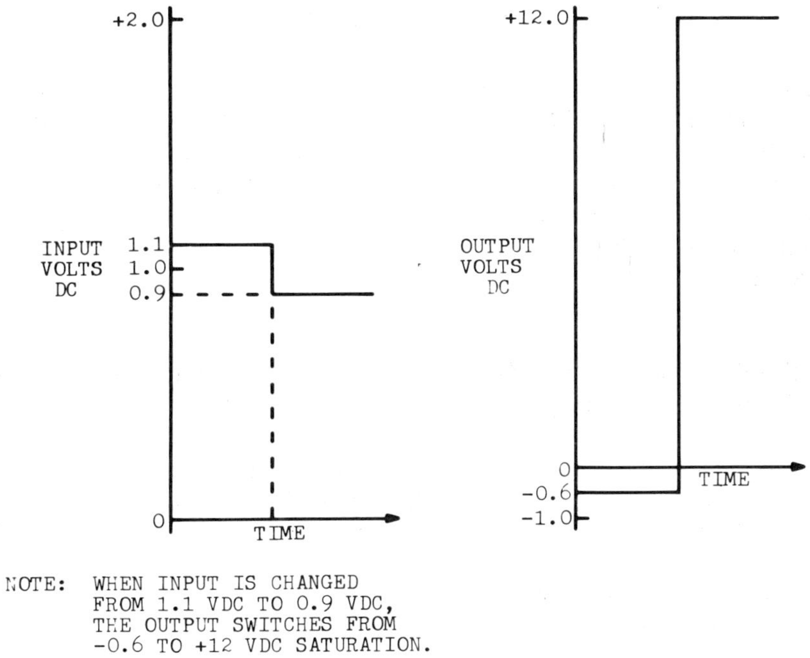


Figure 6-29 Input Versus Output Waveforms for Negative Clamped Simple Comparator

Complex Comparator—Positive Clamped (See figure 6-30) Although current and voltage gain calculations are not necessary for this circuit design, voltage gain should be large so that the open-loop slew rate is acceptable.

Power for the positive clamped comparator is applied to pins 4 and 7. Noninverting pin 3 is grounded. A diode is placed in the feedback loop between output pin 6 and inverting input pin 2. The diode is placed in

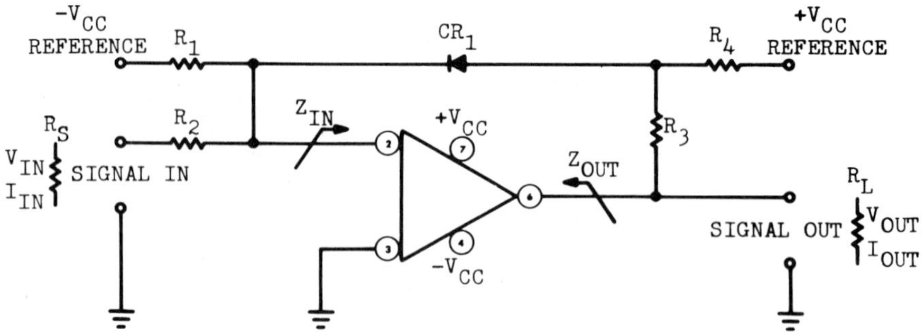


Figure 6-30 Complex Comparator, Positive Clamped

such a manner that the circuit is open-loop for positive input levels and closed-loop for negative input levels. The reference voltage $-V_{CC}$ is compared through resistor $R_1$ with the variable $V_{IN}$ through resistor $R_2$.

When $V_{IN}$ reaches desired clamping levels the output changes state. A wide range of inputs may be used without protection devices for external circuits. The reference voltage $+V_{CC}$ is placed on a voltage divider to pin 6 output and the feedback.

## POSITIVE CLAMPED COMPLEX COMPARATOR—DESIGN PROCEDURES (See figure 6-30)

1. Choose an operational amplifier that will provide the desirable characteristics of the circuit (high open-loop gain).
2. Choose a dc power supply that will be compatible with the op amp.
3. Choose $R_1$ (typically 100 K ohms).
4. Choose $V_{IN}$ (where $V_{IN}$ is input clamping level).
5. Calculate $R_2 = \dfrac{R_1 V_{IN} \text{ (clamping level)}}{-V_{CC} \text{ ref}}$
6. Choose $R_4$ (typically equal to $R_2$).
7. Choose $V_{OUT}$ clamp level (anywhere from 0.6 to 12 volts).
8. Calculate $R_3 = \dfrac{R_4 V_{OUT} \text{ (clamping level)}}{+V_{CC} \text{ ref}}$

therefore: $V_{OUT} = -V_{CC} + 3$ volts (or saturation voltage) when $V_{IN}$ is greater than clamping level

$V_{OUT} =$ chosen output clamping level when $V_{IN}$ is less than input clamping level

9. Choose diode $CR_1$ to withstand feedback current (minor).
10. Connect circuit and power supply on a breadboard.
11. Remove $\pm V_{CC}$ reference, null $V_{IN}$ signals and adjust offset potentiometer for a null at the output.
12. Apply $\pm V_{CC}$ reference and vdc above clamping level to the inverting input.
13. Monitor output with a dc voltmeter.
14. Apply $V_{IN}$ at less than the clamping level.
15. Monitor output with a dc voltmeter.

## POSITIVE CLAMPED COMPLEX COMPARATOR—DESIGN EXAMPLE (See figure 6-30)

**Given:**

$$V_{IN} = 2 \text{ vdc (clamping level)} \qquad R_L = 2 \text{ K}$$

$$V_{OUT} = 8 \text{ vdc (clamping level)}$$

**Procedures:**

1. Choose $\mu$A741 op amp.
2. Choose $\pm 15$ vdc power supply.
3. Choose $R_1 = 100$ K (typical).
4. Choose $V_{IN} = 2$ vdc (clamping level).
5. Calculate $R_2$ :

$$R_2 = \frac{R_1 V_{IN} \text{ (clamping level)}}{V_{CC} \text{ ref}}$$

$$R_2 = \frac{100 \text{ K (2 vdc)}}{15 \text{ vdc}} \cong 13.3 \text{ K}$$

6. Choose $R_4 = R_2 = 13.3$ K.
7. Choose $V_{OUT} = 8$ vdc.
8. Calculate $R_3$ :

$$R_3 = \frac{R_4\ V_{OUT} \text{ (clamping level)}}{+V_{CC} \text{ ref}}$$

$$R_3 = \frac{13.3 \text{ K (8 vdc)}}{15} = 7.1 \text{ K}$$

9. Choose 1N4001 for $CR_1$.

10. Connect circuit and power supply on a breadboard.
11. Remove $\pm V_{CC}$ reference, null $V_{IN}$ signals, and adjust offset potentiometer for a null at the output.
12. Apply $\pm V_{CC}$ reference ($\pm 15$ vdc) and $V_{IN}$ of 2.1 vdc to input.
13. Monitor output for $-12$ vdc (saturation voltage) on voltmeter.
14. Apply $V_{IN}$ of 1.9 vdc.
15. Monitor output of 8 vdc on voltmeter.

**Note:** See figure 6-31 for input versus output waveforms.

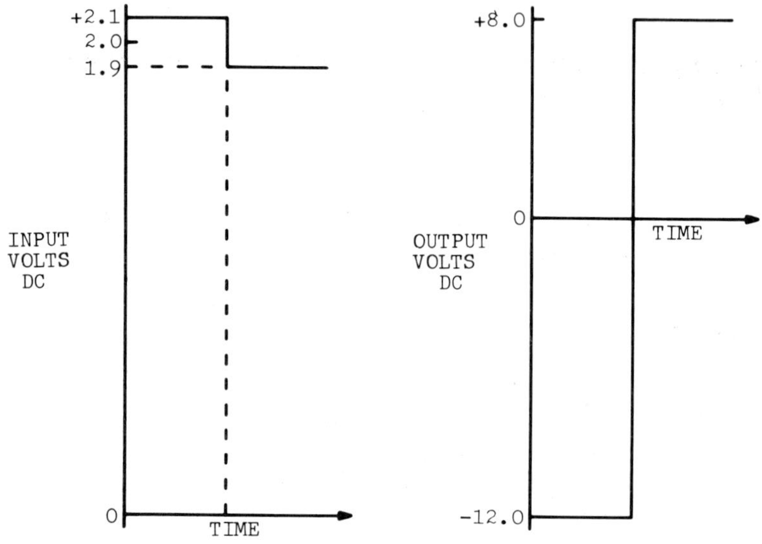


NOTE: WHEN INPUT IS CHANGED FROM 2.1 VDC TO 1.9 VDC, THE OUTPUT SWITCHES FROM -12.0 VDC TO +8.0 VDC.

Figure 6-31 Input Versus Output Waveforms for Postive Clamped Comparator

Complex Comparator—Negative Clamped (See figure 6-32) As with other comparator circuit designs using operational amplifiers, current and voltage gain calculations are not necessary because this amplifier is used only as a voltage comparator. However, voltage gain for negative inputs is said to approach infinity, while voltage gain for positive inputs is said to approach zero.

Power for the comparator is applied to pins 4 and 7. Noninverting input pin 3 is grounded. A diode is placed in the feedback loop between

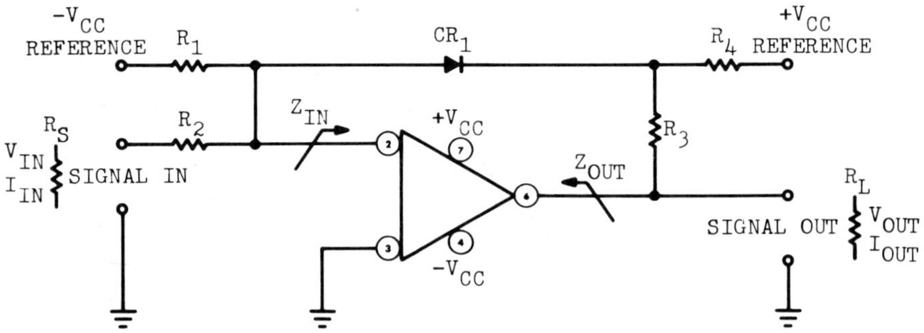


**Figure 6-32** Complex Comparator, Negative Clamped

output pin 6 and inverting input pin 2. The diode is placed in such a manner that the circuit is open-loop for negative input levels and closed-loop for positive input levels. The reference voltage $-V_{CC}$ is compared through resistor $R_1$. The reference voltage $+V_{CC}$ is placed on a voltage divider to output pin 6 and the feedback. Input $V_{IN}$ is applied through resistor $R_2$.

When $V_{IN}$ reaches desired clamping levels, the output changes state. A wide range of inputs may be used without protection for external circuits.

## NEGATIVE CLAMPED COMPLEX COMPARATOR—DESIGN PROCEDURES (See figure 6-32)

1. Choose an operational amplifier that will provide the desirable characteristics of the circuit (high open-loop gain).
2. Choose a dc power supply that will be compatible with the op amp.
3. Choose $R_1$ (typically 100 K ohms).
4. Choose $V_{IN}$ (where $V_{IN}$ is input clamping level).
5. Calculate $R_2 = \dfrac{R_1 V_{IN} \text{ (clamping level)}}{-V_{CC} \text{ ref}}$
6. Choose $R_4$ (typically equal to $R_2$).
7. Choose $V_{OUT}$ clamp level (anywhere from $-0.6$ to $-12$ volts).
8. Calculate $R_3 = \dfrac{R_4 V_{OUT} \text{ (clamping level)}}{+V_{CC} \text{ ref}}$

therefore: $V_{OUT} = V_{CC} - 3$ volts (or saturation voltage) when $V_{IN}$ is less than input clamping level
$V_{OUT}$ = chosen output clamping level when $V_{IN}$ is greater than input clamping level

9. Choose diode $CR_1$ to withstand feedback current (minor).
10. Connect circuit and power supply on a breadboard.
11. Remove $\pm V_{CC}$ reference, null $V_{IN}$ signals, and adjust offset potentiometer for a null at the output.
12. Apply $\pm V_{CC}$ reference and vdc above clamping level to the inverting input.
13. Monitor output with a dc voltmeter.
14. Apply $V_{IN}$ at less than the clamping level.
15. Monitor output with a dc voltmeter.

## NEGATIVE CLAMPED COMPLEX COMPARATOR—DESIGN EXAMPLE (See figure 6-32)

**Given:**

$V_{IN} = 2$ vdc (clamping level) $\qquad R_L = 2$ K

$V_{OUT} = -8$ vdc (clamping level)

**Procedures:**

1. Choose $\mu$A741 op amp.
2. Choose $\pm 15$ vdc power supply.
3. Choose $R_1 = 100$ K.
4. Choose $V_{IN} = 2$ vdc (clamping level).
5. Calculate $R_2$ :

$$R_2 = \frac{R_1 V_{IN} \text{ (clamping level)}}{-V_{CC} \text{ ref}}$$

$$R_2 = \frac{100 \text{ K } (2 \text{ vdc})}{15} \cong 13.3 \text{ K}$$

6. Choose $R_4 = R_2 = 13.3$ K.
7. Choose $V_{OUT} - 8$ vdc.
8. Calculate $R_3$ :

$$R_3 = \frac{R_4 V_{OUT} \text{ (clamping level)}}{+V_{CC} \text{ ref}}$$

$$R_3 = \frac{13.3 \text{ K } (8 \text{ vdc})}{15} \cong 7.1 \text{ K}$$

9. Choose 1N4001 for $CR_1$.
10. Connect circuit and power supply on a breadboard.
11. Remove $\pm V_{CC}$ reference, null $V_{IN}$ signal, and adjust offset potentiometer for a null at the output.
12. Apply $\pm$ reference ($\pm 15$ vdc) and apply $V_{IN}$ of 2.1 vdc to input.
13. Monitor output of $-8$ vdc on voltmeter.
14. Apply $V_{IN}$ of 1.9 vdc.
15. Monitor output of 12 vdc (saturation voltage) on voltmeter.

**Note:** See figure 6-33 for input versus output waveforms.

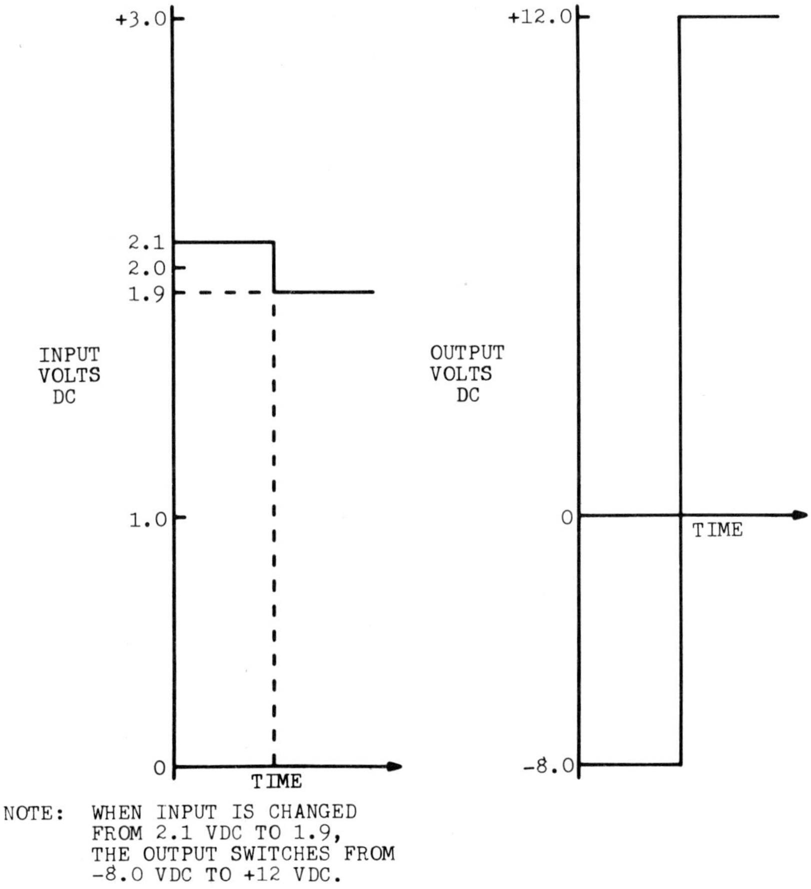


**Figure 6-33** Input Versus Output Waveforms, Negative Clamped Complex Comparator

**Complex Positive and Negative Clamped Comparator** (See figure 6-34) As with other comparator circuit designs, using operational amplifiers, current and voltage gain calculations are not necessary because this amplifier is used only as a voltage comparator. However, the spread of this complex comparator dc level from positive to negative can be greater than a simple clamped comparator because it utilizes both negative and positive feedback.

Power for the clamped comparator is applied to pins 4 and 7. Noninverting pin 3 is grounded. Two diodes are placed in the feedback loop between output pin 6 and inverting pin 2. The diodes are placed in such a manner so that the circuit is closed-loop for positive and negative input levels. Negative reference $-V_{CC}$ is placed across resistor $R_2$ and also resistor $R_6$. Positive reference voltage is placed across $R_4$. The resistors $R_3$ and $R_5$ are used as voltage dividers to the feedback from reference. Input $V_{IN}$ is applied across resistor $R_1$ in comparison to reference.

## COMPLEX POSITIVE AND NEGATIVE CLAMPED COMPARATOR—DESIGN PROCEDURES
(See figure 6-34)

1. Choose an operational amplifier that will provide the desirable characteristics of the circuit (high open-loop gain).
2. Choose a dc power supply that will be compatible with the op amp.

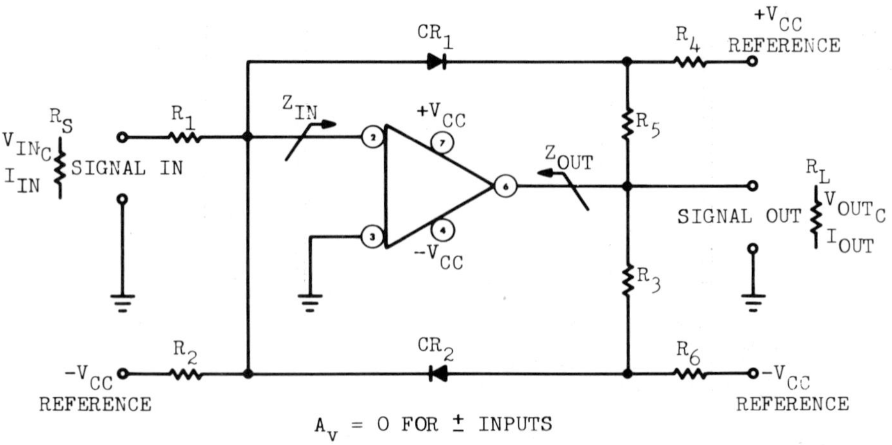


Figure 6-34 Complex Negative and Positive Clamped Comparator

3. Choose $R_2$ (typically 100 K ohms).
4. Choose $V_{IN}$ (where $V_{IN}$ is input clamping level).
5. Calculate $R_1 = \dfrac{R_2 V_{IN} \text{ (clamping level)}}{-V_{CC} \text{ ref}}$.

   (Disregard the minus sign in making this and other resistance calculations.)
6. Choose $R_4$ (typically equal to $R_1$).
7. Choose $V_{OUT}$ clamp level $-0.6$ volt.
8. Calculate $R_5 = \dfrac{R_4 V_{OUT} \text{ (negative clamping level)}}{+V_{CC} \text{ ref}}$.

   (Disregard the minus sign in making this and other resistance calculations.) Therefore $V_{OUT}$ = chosen output clamp level when $V_{IN}$ is greater than input clamp level.
9. Choose $R_6$ (typically equal to $R_1$).
10. Choose $V_{OUT}$ clamp level (8 volts).
11. Calculate $R_3 = \dfrac{R_6 V_{OUT} \text{ (positive clamping level)}}{-V_{CC} \text{ ref}}$.
12. Choose diode $CR_1$ to withstand feedback current (minor).
13. Connect circuit and power supply on a breadboard.
14. Remove $\pm V_{CC}$ reference, null $V_{IN}$ signals, and adjust offset potentiometer for a null at the output.
15. Apply $\pm V_{CC}$ reference and vdc above clamping level to the inverting input.
16. Monitor output with a dc voltmeter.
17. Apply $V_{IN}$ at less than the clamping level.
18. Monitor output with a dc voltmeter.

## COMPLEX POSITIVE AND NEGATIVE CLAMPED COMPARATOR—DESIGN EXAMPLE
(See figure 6-34)

**Given:**

$V_{IN} = 2$ vdc (clamping level) $\qquad R_L = 2$ K

$V_{OUT} = 8$ vdc (clamping level)

$V_{OUT} = -6$ vdc (clamping level)

**Procedures:**

1. Choose $\mu$A741 op amp.
2. Choose $\pm 15$ vdc power supply.
3. Choose $R_2 = 100$ K.
4. Choose $V_{IN} = 2$ vdc (clamping level).
5. Calculate $R_1$:

$$R_1 = \frac{R_2\, V_{IN}\text{ (clamping level)}}{-V_{CC}\text{ ref}}$$

$$R_1 = \frac{100\text{ K (2 vdc)}}{15} = 13.3\text{ K}$$

6. Choose $R_4 = R_1 = 13.3$ K.
7. Choose $V_{OUT}$ clamping level $= -6$ vdc.
8. Calculate $R_5$ :

$$R_5 = \frac{R_4\, V_{OUT}\text{ (negative clamping level)}}{+V_{CC}\text{ ref}}$$

$$R_5 = \frac{13.3\text{ K }(-6\text{ vdc})}{15} = 5.3\text{ K}$$

9. Choose $R_6 = R_1 = 13.3$ K.
10. Choose $V_{OUT}$ clamping level $= 8$ vdc.
11. Calculate $R_3$ :

$$R_3 = \frac{R_6\, V_{OUT}\text{ (positive clamping level)}}{+V_{CC}\text{ ref}}$$

$$R_3 = \frac{13.3\text{ K }(+8\text{ vdc})}{15} = 7.1\text{ K}$$

12. Choose 1N4001 for $CR_1$.
13. Connect circuit and power supply on a breadboard.
14. Remove $\pm V_{CC}$ reference, null $V_{IN}$ signal, and adjust offset potentiometer for a null at the output.
15. Apply $\pm V_{CC}$ reference ($\pm 15$ vdc) and $V_{IN}$ of 2.1 vdc to input.
16. Monitor output for $-6$ vdc on voltmeter.
17. Apply $V_{IN}$ of 1.9 vdc.
18. Monitor output of 8 vdc on voltmeter.

**Note:** See figure 6-35 for input versus output waveforms.

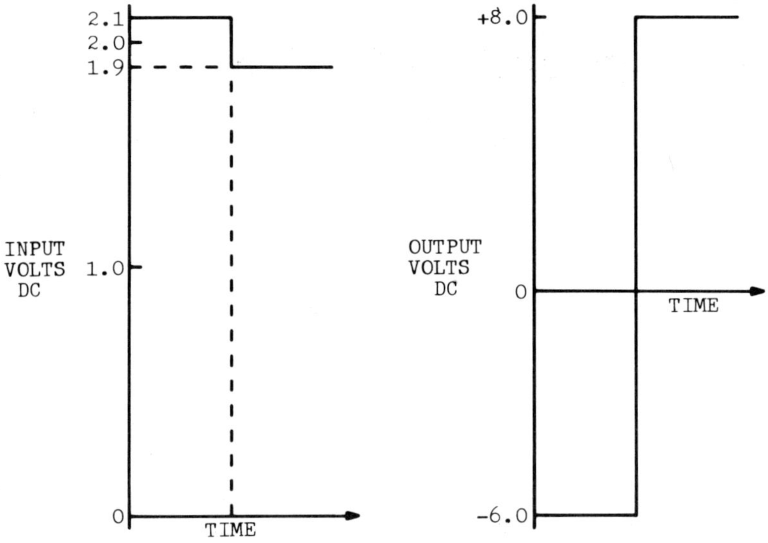


NOTE: WHEN INPUT CHANGES FROM 2.1 VDC TO 1.9 VDC, THE OUTPUT SWITCHES FROM -6 VDC TO +8 VDC.

**Figure 6-35** Input Versus Output Waveforms for Complex Negative and Positive Clamped Comparator

# 7
# The Bridge and the Op Amp

In all probability the resistance bridge is the most elementary and yet the most used circuit in instrumentation and related fields of measurement. In fact the resistance bridge is used throughout industry in control systems. The prominent and basic type of bridge is the Wheatstone bridge. The Wheatstone bridge was announced by Sir Charles Wheatstone, a prominent British physicist and came into use around 1843. There is some controversy on whether the circuit was really a Wheatstone creation. Some historians give credit to Samuel Christie, another Britisher, as early as 1833. Whatever the case may be, the Wheatstone bridge was the first device/circuit that accurately measured resistance. The elementary Wheatstone bridge is used to determine resistance of an unknown value. It accomplishes this by comparing other known and precision resistors with the one under question. The original Wheatstone bridge consisted of three known resistances, a battery and a galvanometer. The unknown resistance was placed in a voltage divider with one of the known resistors (variable). The other two resistors were also set up as a fixed voltage divider. The galvanometer was centered between the two series voltage dividers. Adjustment was then made with the variable resistance until the galvanometer indicated a null (no current flow). Once a null was achieved, simple mathematical proportion was used to calculate the value of the unknown resistance.

## THE TRANSDUCER

The transducer is a device that converts energy from one form to another. In the instrument field it is generally understood that the transducer converts a measurand into an electrical quantity. The measurand is something that is to be measured; it can be energy or it can be applicable to nonenergy situations. For instance, the measurand could be the pressure of gas within a chamber; or, it could be the monitoring of a distance from one point to another.

Basically there are two types of transducer: the active and the passive transducer. The active transducer is one that generates a voltage direct from the energy-producing source. An example of the active transducer is the acceleration device. The accelerometer converts mechanical energy to an electrical signal. The passive transducer is one that changes an electrical characteristic such as resistance. An example of the passive transducer is the thermistor. Temperature changes the resistance and the resistance, in turn, causes a change in current flow in the circuit involved.

The transducer is extremely well suited for use in a bridge circuit. And, of course, the bridge is extremely well suited for use with an operational amplifier.

The importance of a bridge is that a very small amount of signal can produce a relatively large signal to drive a sensitive ammeter (galvanometer). Output signals are made more effective by the connection of an amplifier in the output of the bridge in place of a galvanometer.

The amplifier is usually a differential amplifier which compares the difference voltage between two points. The operational amplifier represents the state of the art in differential amplifiers. It has a very high input impedance and has evolved into instrumentation amplifiers that have the ability to prevent noise from being amplified.

**The Voltage Divider (See figure 7-1)** Figure 7-1 illustrates a basic signal conditioner, the forerunner of the bridge network. The purpose of this circuit is to monitor a change in the resistance $R_2$. The voltage $E_{R2}$ is a measurement of the voltage drop across $R_2$. This, of course, is a steady-state component. A change in the resistance $R_2$ will present a change in the voltage $E_{R2}$. The capacitor $C_1$ is charged to the steady-state voltage of $E_{R2}$ and passes only a change to the output $E_{OUT}$. This change is usually passed on to a meter for display. The scale of the meter is selected to monitor the amplitude of change and therefore is not affected by the dc voltage across $R_2$. The static voltage level of the resistor $R_2$ is calculated using the simple voltage divider method. This calculation is as follows:

$$E_{R2} = V_{DC} \frac{R_2}{R_1 + R_2}$$

$R_2/(R_1 + R_2)$ is a ratio of $R_2$ against the total resistance in the voltage divider $R_1 + R_2$.

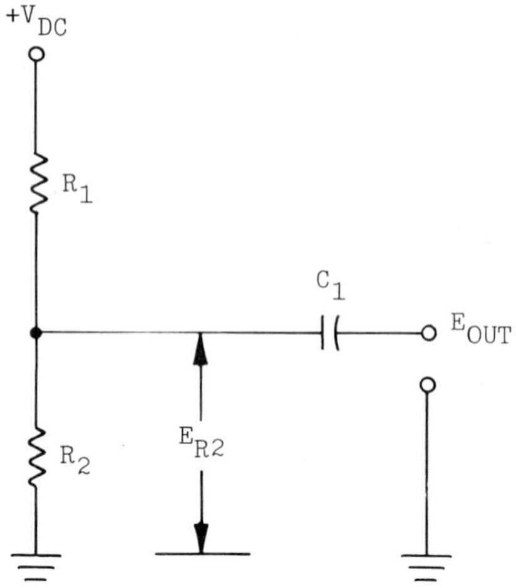


Figure 7-1 A Most Basic Signal Conditioner

## BRIDGE TYPES

The elementary measurement circuit used in instrumentation and specifically with strain gages is the resistance bridge. The prominent circuit type is the Wheatstone bridge. The elementary Wheatstone bridge is used to determine the resistance of any unknown resistance. This same bridge is used extensively in automatic control systems and also with electrical measuring instruments.

**The Basic Wheatstone Bridge** (See figure 7-2) $R_1$ and $R_2$ are a voltage divider from the power source. $R_{UNK}$ (the unknown resistance) and $R_{VAR}$ (variable calibrated resistance) represent the second voltage divider from the power source. The $R_1$, $R_2$ and $R_{VAR}$ resistances are all very accurate components. Their values are known.

A precision galvanometer is placed between points A and B. When power is applied to the circuit, the galvanometer will indicate current flow in either the point A or point B direction. Resistor $R_{VAR}$ is then varied until the galvanometer indicates no current flow. This shows that the bridge is balanced. When it is balanced, the bridge sets up a mathematical voltage ratio between the two voltage dividers. This ratio is as follows:

$$\frac{I_1 R_1}{I_1 R_2} = \frac{I_2 R_{UNK}}{I_2 R_{VAR}}$$

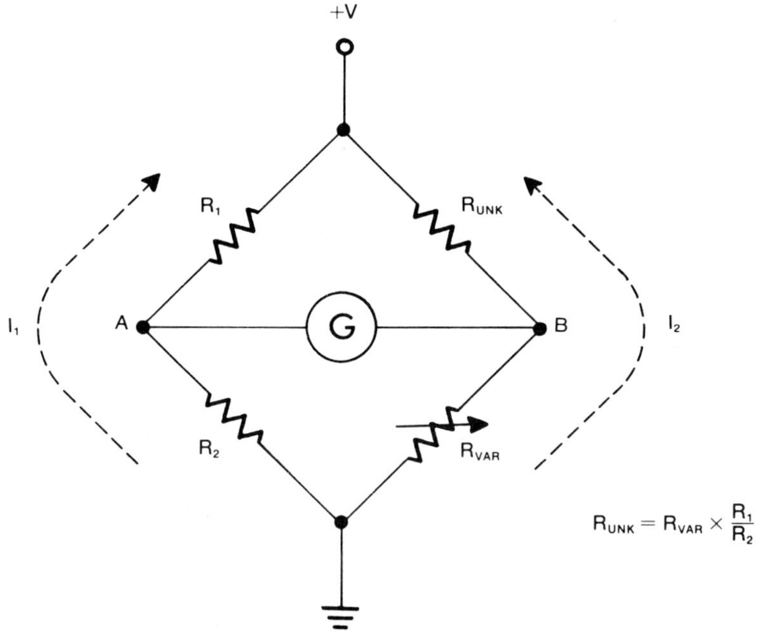


Figure 7-2 A Wheatstone Bridge

The current component can be removed from the equation because current is equal in each leg. The ratio then reads:

$$\frac{R_1}{R_2} = \frac{R_{\text{UNK}}}{R_{\text{VAR}}} \quad \text{or} \quad R_{\text{UNK}} = R_{\text{VAR}} \times \frac{R_1}{R_2}$$

**The Wheatstone Bridge Used in Signal Measurement** (See figure 7-3) In the search for accuracy, the Wheatstone bridge seems to stand tall for measurement of small resistance changes. Figure 7-3 is representative of a balanced bridge circuit. The terminals of the output ($E_0$) are in balance and have a difference of potential of zero. The output $E_0$ is zero.

A bridge is used for three reasons: as a circuit for determining an unknown resistance, as a null detector, and as a circuit to measure the difference between two voltages or current.

There is a common resistance ratio for the bridge shown in the figure 7-3. This ratio is

$$\frac{R_1}{R_4} = \frac{R_2}{R_3}$$

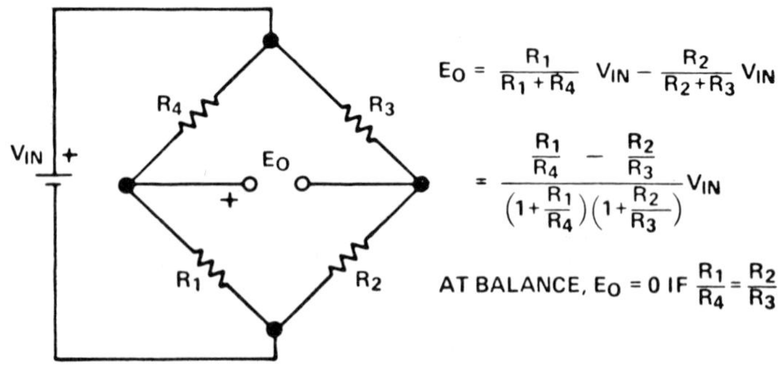


Figure 7-3 Basic Bridge Circuit (Courtesy, Analog Devices)

If three resistances of this ratio are known, the other can easily be calculated. For instance if the resistance $R_3$ is unknown, then

$$R_3 = \frac{R_2 R_4}{R_1}$$

As a null detector, the output of the bridge would be at null ($E_0 =$ 0 volts) when and if the bridge were balanced.

As a detector of voltage the input $V_{IN}$ is placed between opposite terminals, while the output $E_0$ is taken from a second set of terminals. In essence, the bridge output is measuring the difference voltage between the leg pairs. The first leg is a series voltage divider looking at the voltage across resistor $R_1$.

$$V_{R1} = \frac{R_1}{R_1 + R_4} V_{IN}$$

The second leg is a series voltage divider looking at the voltage across resistor $R_2$.

$$V_{R2} = \frac{R_2}{R_2 + R_3} V_{IN}$$

The calculation of $E_0$ is the difference voltage between the two series legs.

$$E_0 = \frac{R_1}{R_1 + R_4} V_{IN} - \frac{R_2}{R_2 + R_3} V_{IN}$$

or
$$E_0 = \frac{\dfrac{R_1}{R_4} - \dfrac{R_2}{R_3}}{\left(1 + \dfrac{R_1}{R_4}\right)\left(1 + \dfrac{R_2}{R_3}\right) V_{IN}}$$

The bridge as shown in the illustration in figure 7-3, allows for the detection and measurement of small resistance changes. A change from bridge balance may easily be monitored. The output $E_0$ changes as any adjacent leg resistance changes. The change is represented by a difference of potential between the parallel legs (at $E_0$). There also may be cases when adjacent legs change resistance by the same amount. In this event, the bridge would still be balanced $R_1/R_4 = R_2/R_3$. The output voltage ($E_0$) would remain at a null of zero.

In transducer arrangements, the ratio of fixed resistances such as $R_2/R_3$ is called ratio K.

Let's assume that $R_2/R_3$ are fixed at a constant ratio K. If $R_1$ is unknown, and $R_4$ is an accurate variable resistance, then the size of $R_1$ can be determined by adjusting $R_4$ until a null is achieved. The null will occur when $R_1/R_4 = R_2/R_3$. If $R_4$ is fixed then a null will occur when $R_1$ becomes equal to K times $R_4$.

**The Fully Active Full Bridge Circuit** A fully active full bridge is one in which all resistances in the bridge (four) have identical resistances. The four bridge resistances also have the same value of change ($\Delta R$). The output of this bridge is maximum because all bridge resistances are the same and all respond to the same amount of change. There is minimum effect because of temperature or other environmental activity. The bridge is usually enclosed while an external circuit is used for measurement. The formula for defining this bridge is as follows:

$$\frac{E_0}{V_{IN}} = \frac{\Delta R}{R}$$

where
$$R_{IN} = R_{OUT} = R$$
$$V_{IN} = \text{applied voltage}$$

The $\Delta R$ in the equation represents a change in resistance because of a changing load.

**Bridge Circuit with a Variable Element** For most transducer and instrument circuits that utilize bridge circuits, one or more of the bridge resistors vary. In figure 7-4, a bridge circuit is illustrated that has a single variable element. All resistors in the bridge are the same value with the exception of $R_1$, which is variable. $R_1$ is variable by a factor of $1 + X$.

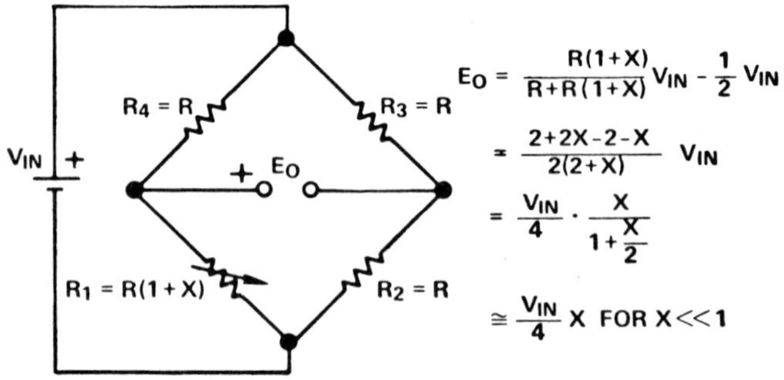


Figure 7-4 Bridge Circuit with a Variable Element (Courtesy, Analog Devices)

The voltage output $E_0$ equation is written in the same general manner as the basic bridge. For small change of $X$ this circuit is linear.

The ratio of the fixed resistance leg is

$$\frac{R}{R+R} \quad \text{or} \quad \frac{1}{2}$$

The ratio of the variable resistance leg is

$$\frac{R(1+X)}{R+R(1+X)}$$

The output of this bridge ($E_0$) is the difference of potential between the voltage divided legs.

$$E_0 = \frac{R(1+X)}{R+R(1+X)} V_{IN} - \frac{R}{R+R} V_{IN}$$

or

$$E_0 = \frac{R(1+X)}{R+R(1+X)} V_{IN} - \tfrac{1}{2} V_{IN}$$

or

$$E_0 = \frac{2+2X-2-X}{2(2+X)} V_{IN}$$

or

$$E_0 = \frac{V_{IN}}{4} \cdot \frac{X}{1+\dfrac{X}{2}}$$

or

$$E_0 \cong \frac{V_{IN}}{4} X \quad \text{for} \quad X \ll 1$$

The bridge just discussed is often called the half-active half-bridge. It is represented partially by a pair of inactive resistors $R_2$ and $R_3$. If used as a transducer, these resistors may be placed outside the transducer case. The other two resistors $R_1$ and $R_4$ are identical—one active ($R_1$) and one inactive ($R_4$). With only one active leg the bridge is a simple transducer.

**Bridge Circuit with Two Variable Elements** The bridge that is illustrated in figure 7-5 has two variable elements, one on each voltage divider leg. In the figure, these variable resistances are opposite from each other. The variable resistances $R(1+X)$ are variable by the same quantity. The fixed resistances ($R$) are equal to each other. The voltage output $E_0$ equation is written in the same general manner as the basic bridge.

The ratio of the left leg in the bridge in the illustration is

$$\frac{R(1+X)}{R+R(1+X)}$$

The output of this bridge ($E_0$) as with previous bridges is the difference of potential between the two voltage divided legs.

$$E_0 = \frac{R(1+X)}{R+R(1+X)} V_{\text{IN}} - \frac{R}{R+R(1+X)} V_{\text{IN}}$$

or

$$E_0 = \left[\frac{R(1+X)}{R+R(1+X)} - \frac{R}{R+R(1+X)}\right] V_{\text{IN}}$$

or

$$E_0 = \frac{X}{2+X} V_{\text{IN}}$$

or

$$E_0 = \frac{V_{\text{IN}}}{2} \cdot \frac{X}{1+\frac{X}{2}}$$

or

$$E_0 \cong \frac{V_{\text{IN}}}{2} X \quad \text{for} \quad X \ll 1$$

This bridge is used basically to increase the sensitivity of the circuit. Sensitivity is a ratio of excitation voltage to the maximum expected change in the output $E_0$. Note that the output and the sensitivity are double that of a single variable bridge. This bridge, if utilized as a transducer may have the inactive legs located externally.

A variation of the circuit in figure 7-5 is one in which the two

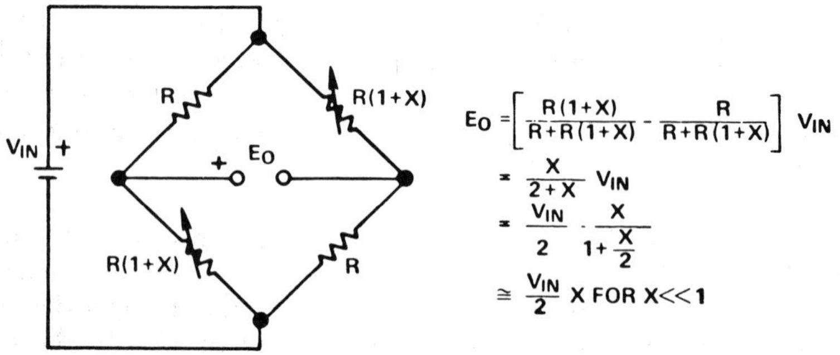


**Figure 7-5** Bridge Circuit with Two Variable Elements (Courtesy, Analog Devices)

variables are equal but positive and negative. The equation for defining this bridge is as follows:

$$\frac{E_0}{V_{IN}} = \frac{\Delta R}{2R}$$

**Bridge Circuit with Four Variable Elements** The bridge that is illustrated in figure 7-6 has four variable elements. Two opposite resistances are variable $R(1 + X)$ while the other two opposite resistances are variable $R(1 - X)$. The change is equal and increases/decreases occur at the same ratio. The output $E_0$ of this bridge is four times that of the single variable bridge which provides a more linear, more sensitive output.

The ratio of the left leg in the illustration is

$$\frac{R(1 + X)}{R(1 - X) + R(1 + X)}$$

or

$$\frac{R(1 + X)}{2R}$$

The ratio of right leg of the bridge in the illustration is

$$\frac{R(1 - X)}{R(1 + X) + R(1 - X)}$$

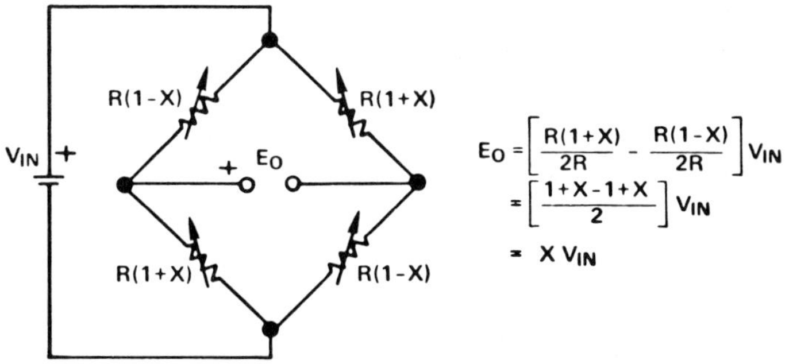


Figure 7-6 Bridge Circuit with All Elements Variable (Courtesy, Analog Devices)

or

$$\frac{R(1 - X)}{2R}$$

The output of this bridge ($E_0$) as with previous bridges is the difference of potential between the two voltage divided legs as in the following equation

$$E_0 = \frac{R(1 + X)}{2R} V_{IN} - \frac{R(1 - X)}{2R} V_{IN}$$

or

$$E_0 = \left[\frac{R(1 + X)}{2R} - \frac{R(1 - X)}{2R}\right] V_{IN}$$

or

$$E_0 = \left[\frac{1 + X - 1 + X}{2}\right] V_{IN}$$

or

$$E_0 = X V_{IN}$$

As with the two variable element bridges, this bridge is much more sensitive than single element bridges.

**Bridge Circuit with a Potentiometer** In the figure 7-7, a bridge is shown with two equal resistances on one branch leg and a zero center potentiometer on the other leg. The potentiometer takes the place of two adjacent resistances. The wiper of the potentiometer represents the measurand. Since the potentiometer wiper will move in either direction (the change being $1 - X$ or $1 + X$) the output will be twice that of the

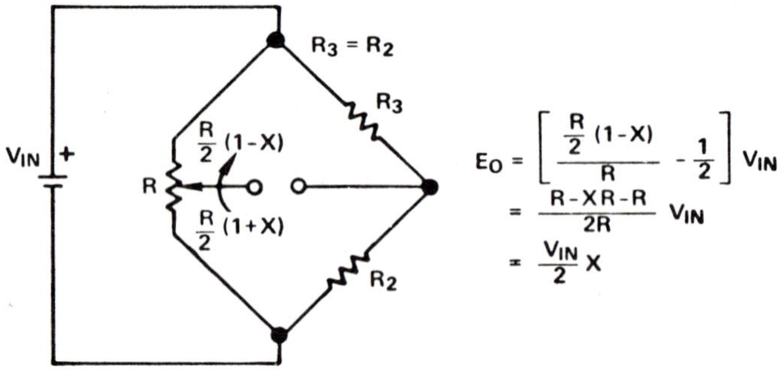


Figure 7-7 Bridge Circuit with a Potentiometer (Courtesy, Analog Devices)

single variable bridge. The voltage output is determined by the following equation:

$$E_0 = \frac{\frac{R}{2}(1 - X)}{R} V_{IN} - \tfrac{1}{2}V_{IN}$$

or

$$E_0 = \left[\frac{\frac{R}{2}(1 - X)}{R} - \frac{1}{2}\right] V_{IN}$$

or

$$E_0 = \frac{R - XR - R}{2R} V_{IN}$$

or

$$E_0 = \frac{V_{IN}}{2} X$$

Although this circuit is expected to be linear, it may have the nonlinearity that is axiomatic of the potentiometer. This bridge and others can be made more linear by making the ratios on adjacent legs larger. This action must be taken by sacrificing sensitivity.

**An Active Bridge with an Op Amp in Series** The illustration in figure 7-8 is an active bridge with an op amp placed in series with the variable arm $R(1 + X)$. The op amp adds a voltage in series with the variable arm $R(1 + X)$ that is equal and opposite in polarity to the voltage across the resistance $R(1 + X)$. The op amp has an extremely low output impedance (less than 25 Ω). Therefore the bridge can be measured as a low impedance. The circuit is linear and has twice the gain of a single element bridge.

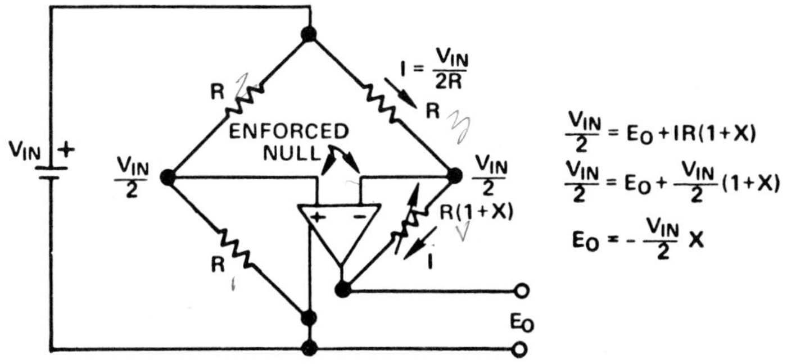


Figure 7-8 An Active Bridge (Courtesy, Analog Devices)

Input to the op amp is $V_{IN}/2$. This input is equal to $E_0$ plus the voltage felt on the variable arm $IR(1 + X)$.

$$\frac{V_{IN}}{2} = E_0 + IR(1 + X)$$

substituting

$$\frac{V_{IN}}{2} = E_0 + \frac{V_{IN}}{2}(1 + X)$$

then

$$-E_0 = \frac{V_{IN}}{2} + \frac{V_{IN}}{2}X - \frac{V_{IN}}{2}$$

and

$$E_0 = -\frac{V_{IN}}{2}X$$

Note that the resistances $R$ are all equal which makes the voltage drop at the voltage divider centers equal to $V_{IN}/2$. The current flow $I$ through each leg is equal to $V_{IN}/2R$.

**Bridge Excitation** The designer must use his own discretion on the choice of excitation for his bridge. The decision is based on what is available, the peculiarities of his design and the precision required. One basic method is to use a voltage reference such as the Analog Devices Model AD581 along with an op amp and a current booster. This provides a regulated voltage on the bridge input. (See figure 7-9.) Note that the 10 volts applied to 250 Ω bridge provides 40 mA of current flow.

A second excitation system, shown in figure 7-10, uses a system approach where a signal conditioning power supply is utilized. This system ensures that a precise voltage is maintained on the terminals of the bridge. Sensing leads are provided that automatically adjust for voltage changes due to leads.

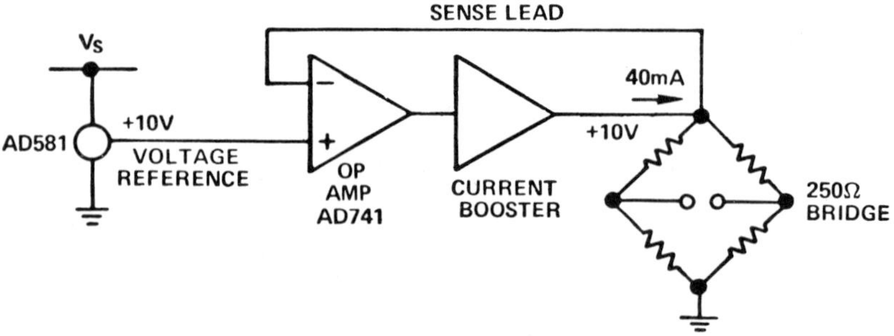


Figure 7-9 Basic Bridge Drive Circuit (Courtesy, Analog Devices)

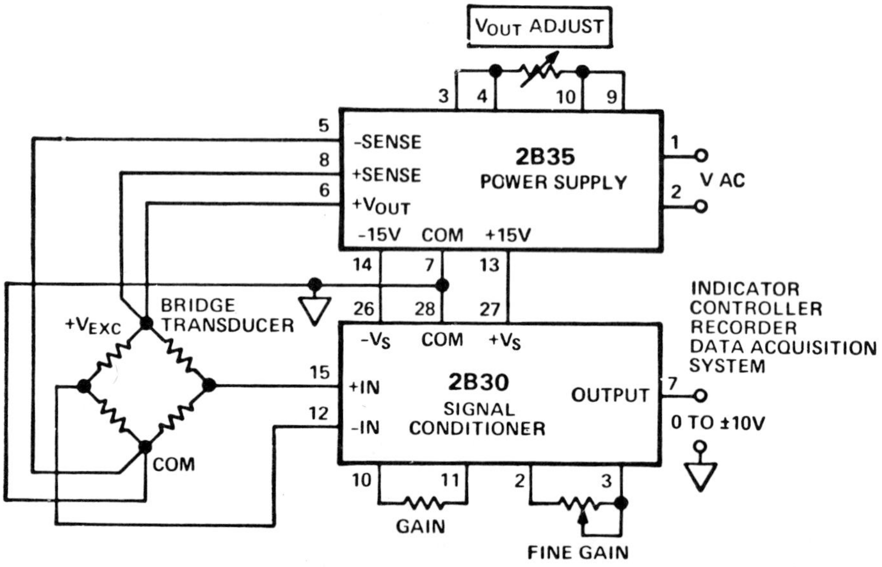


Figure 7-10 A Transducer Power Supply Used as a Bridge Drive (Courtesy, Analog Devices)

A third system of bridge excitation is shown in figure 7-11. This system is called the ratiometric system and is used for high precision measurements. A supply voltage reference is provided by the Analog Devices Model AD580 along with an op amp and a current booster. The same reference is used for the readout and the power supply. In this case the readout is a digital panel meter.

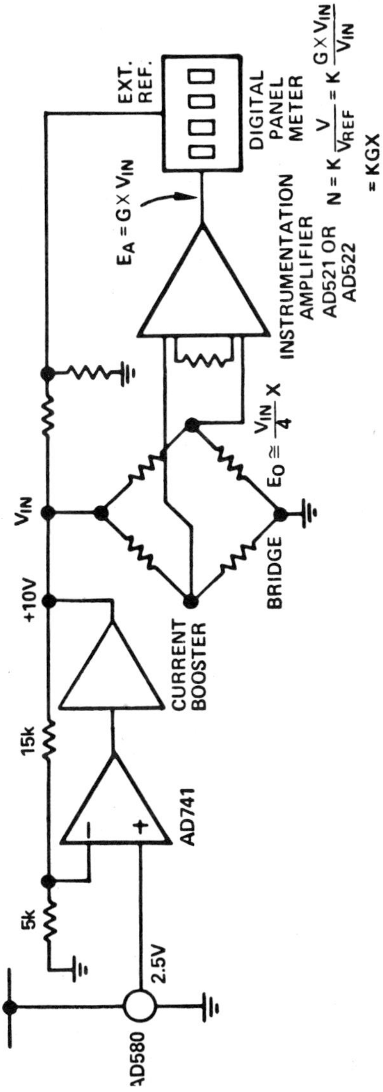


Figure 7-11 The Ratiometric Excitation System (Courtesy, Analog Devices)

## THE OP AMP USED AS A BRIDGE AMPLIFIER
(See figure 7-12)

The usefulness of the op amp as a bridge amplifier is most desirable. The costs may be high depending on your particular application and whether you are building one or a thousand circuits. The simple nature of this circuit makes it worthwhile.

The circuit is excited by some predetermined $V_{IN}$. The resistances $R$ are equal with one variable $R(1 + X)$. One leg of the bridge is fed to the inverting input of the op amp. The variable leg is fed to the noninverting input. The feedback resistance $R_1$ is coupled between the output of the op amp and the inverting input. A second resistance $R_1$ of equal value to the feedback resistance $R_1$ is inserted between the noninverting input and ground.

Voltage gain is equal to

$$A_v = \frac{E_0}{V_{IN}}$$

Input voltage to the inverting input is the voltage divider,

$$\frac{R}{2R} V_{IN}$$

Input voltage to the noninverting input is the voltage divider,

$$\frac{R(1 + X)}{R + R(1 + X)} V_{IN}$$

The circuit has voltage gain and the op amp does provide a null

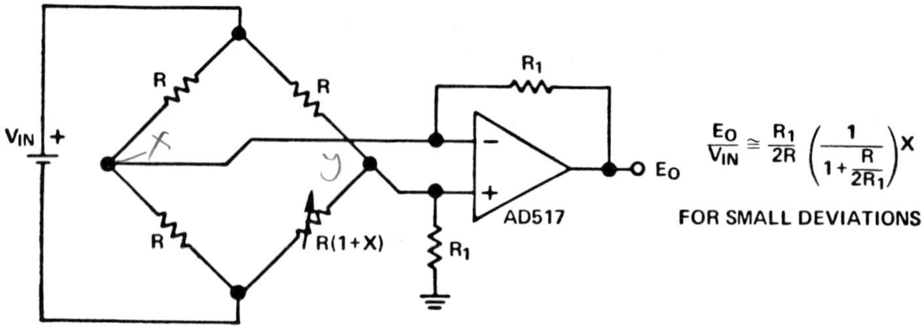


Figure 7-12 Operational Amplifier as a Bridge Amplifier (Courtesy, Analog Devices)

across the bridge. In order to achieve an acceptable common mode rejection (CMR) the external resistors must be matched. However, the circuit may still be more nonlinear than the bridge circuit itself.

Output of the bridge and the amplifier is calculated by the following equation.

$$\frac{E_0}{V_{IN}} \cong \frac{R_1}{2R}\left(\frac{1}{1+\frac{R}{2R_1}}\right)X$$

The circuit is used for small deviations from bridge null.

## CONNECTING A BRIDGE AND AN OP AMP (See figure 7-13)

The illustration in figure 7-13A is typical of a basic bridge circuit with a single variable $R_X$ representing a transducer. A reference voltage $V_{REF}$ is placed between the two leg pairs. The output is taken from each pair's center connection. This circuit is representative of a normal transducer differential signal interconnection. The outputs are $R_{OUT_{HIGH}}$ and $R_{OUT_{LOW}}$.

Resistance outputs are calculated in the following manner:

$$R_{OUT_H} = \frac{R}{2}$$

and

$$R_{OUT_L} = \frac{R(R_X)}{R+R_X}$$

The reader will note that the equations cannot include noise and lead resistance. Therefore the differential inputs are not balanced and the output impedances presented to the operational amplifier are not exactly zero.

In figure 7-13B, an operational amplifier is shown in differential connection. This circuit has low input impedance. The two inputs as just discussed, relative to the bridge in figure 7-13A, are not balanced. Common mode rejction (CMR) to eliminate the noise is dependent on precise external resistor matching.

Gain can be balanced by the following equation.

$$A_v = \frac{R_2}{R_1} = \frac{R_2'}{R_1'}$$

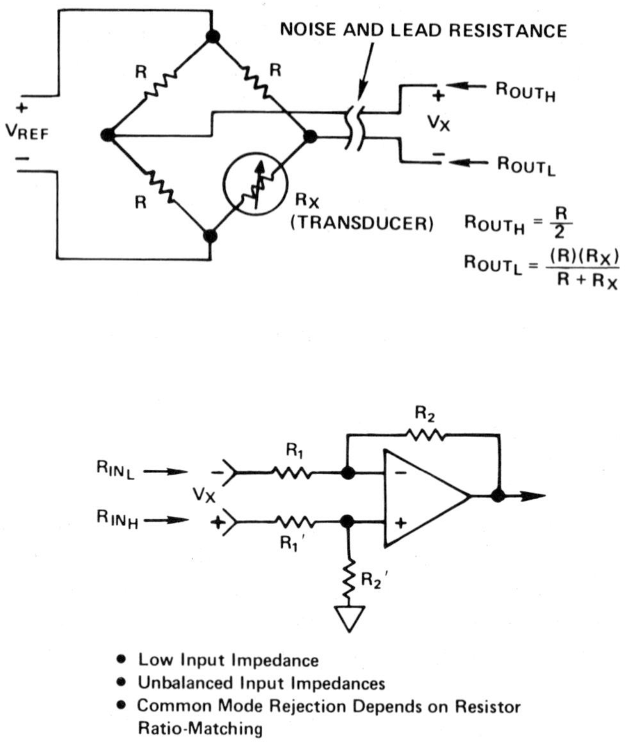


Figure 7-13 Connecting a Bridge and an Op Amp (Courtesy, Analog Devices)

Input impedance is balanced by the following equation.

$$R_1 = R_1' + R_2'$$

In order to provide balanced impedance paths for bias currents to avoid offset drift, the following must be equated.

$$\frac{(R_1)(R_2)}{R_1 + R_2} = \frac{(R_1')(R_2')}{R_1' + R_2'}$$

These conditions are exclusive to the ideal transducer with lead resistance still being a problem.

Most practical sensor circuitry requires differential connections. The impedance of a bridge and its leads is nonzero and it is easy to pick up noise. It is for these reasons that the instrumentation amplifier was developed and is in use today. The instrumentation amplifier is used with a bridge (and other circuitry) when the impedance and noise are such that it is difficult to acquire a useful signal. The instrument amplifier is discussed in the next chapter.

## CHOPPER-STABILIZED BRIDGE AMPLIFIER
(See figure 7-14)

The chopper operational amplifier is used with the bridge when the designer wishes low drift with time and temperature. The chopper is a noninverting amplifier that produces a dc output that is offset corrected. The input and the output voltage on these devices share the same terminal. The power supply floats and therefore no common mode voltage (noise error) appears at the amplifier. Furthermore the drift is very low. As you can understand, since the power supply must be extremely stable and regulated, there will be higher than usual costs. Gain calculation for this circuit is as follows:

$$A_v(G) = 1 + \frac{R_2}{R_1}$$

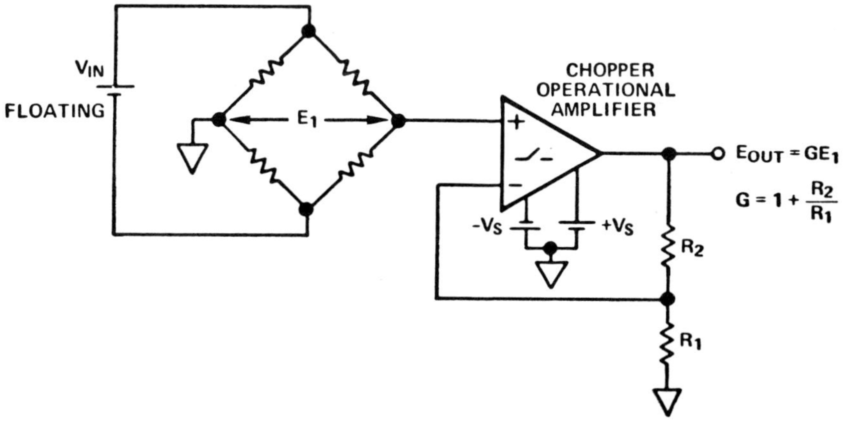


Figure 7-14 Chopper-Stabilized Bridge Amplifier (Courtesy, Analog Devices)

## CHOPPER-STABILIZED BRIDGE AMPLIFIER WITH FLOATING SUPPLY (See figure 7-15)

In this circuit a chopper amplifier is used as a freely floating preamp to an isolation amplifier. Power for the chopper is derived from the isolation amplifier. All three components—the bridge, chopper, and the isolation amp—utilize the same floating ground. The excitation, the bridge, and the output section of the isolation amplifier are also grounded to common. This circuit provides extremely high accuracy.

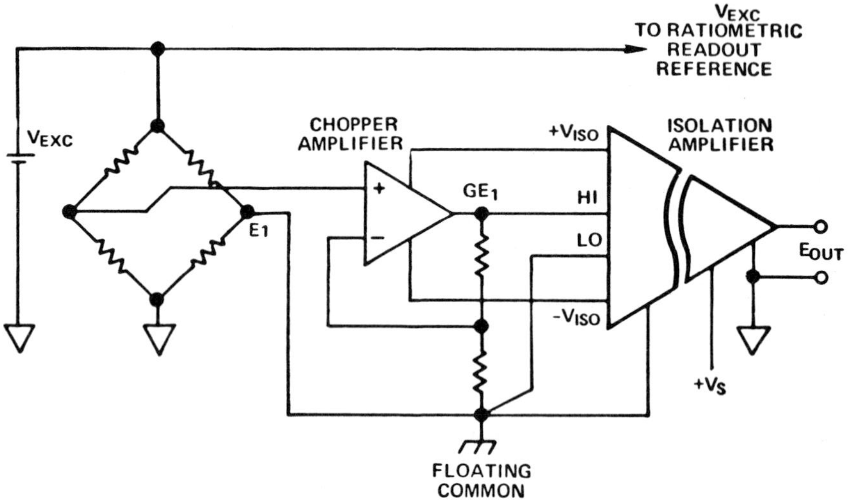


**Figure 7-15** Chopper-Stabilized Bridge Amplifier with Floating Supply Voltage (Courtesy, Analog Devices)

# 8
# Instrumentation Amplifiers

Transducers are not usually electrically cooperative. The transducer user is plagued by nonlinear devices, electrical interference, high output impedances, inconvenient voltage ranges, and unbalanced outputs. Furthermore, the transducer is subjected to noise and temperature changes, and often works in remote locations. The use of the op amp for a wide variety of applications has been a boon to the industry. Costs of general-purpose devices are very low, with increases only for speed and precision. The general-purpose device is versatile and can do other things than simply amplify. The transducer, however, with its special problems, has created the need for something better. Use of the op amp for all applications may not be the answer.

The practical transducer shown in figure 8-1A uses differential connections to a bridge. The output impedance of the bridge is not zero and is unbalanced. Long leads add more resistance and noise pickup is hard to avoid. In a differential amplifier such as shown in figure 8-1B,

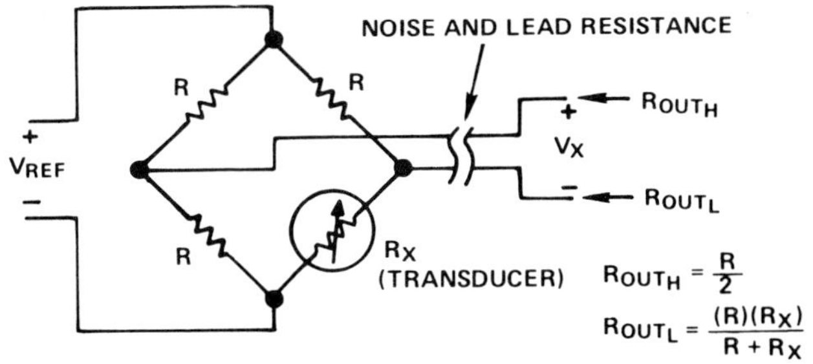


(a)

Figure 8-1 Operational Amplifier Limitations (Courtesy, Analog Devices)

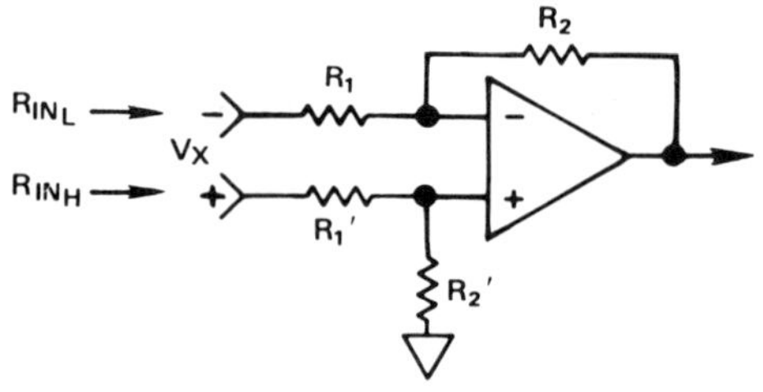


Figure 8-1 *continued*

common-mode rejection (CMR) is dependent on the ratio matching of the external resistances. The op amp configuration may not be able to reject the common-mode noise. It is for this reason that the instrumentation amplifier was created. The instrumentation amplifier has the qualities required for bridge and transducer applications, high input impedance, high gain, and uncommonly high common-mode rejection (CMR). The only tradeoff is that the gain of the instrumentation amplifier is held within a specified range. Gain values are usually under 1,000. This is really not a problem, however. If better gain, control, and quantity are required, a second op amp may be added at the output.

## CHARACTERISTICS OF THE INSTRUMENTATION AMPLIFIER (See figure 8-2)

An instrumentation amplifier is a high-CMR amplifier with high input impedance. It is committed to a fixed gain block. The amplifier measures the difference between two voltages placed at input terminals and amplifies it (precisely).

The output is taken from a pair of external terminals. If the inputs ( + and – ) are equal, the output will be zero. Gain is set by the gain resistance $R_G$ and trimmed (in some instrumentation amplifiers) by the gain trim pot $R_S$. Offset may be trimmed if necessary.

The typical instrumentation amplifier has the following characteristics:

1. High input impedance
2. Low bias current
3. High common-mode rejection (CMR)
4. Balanced differential inputs

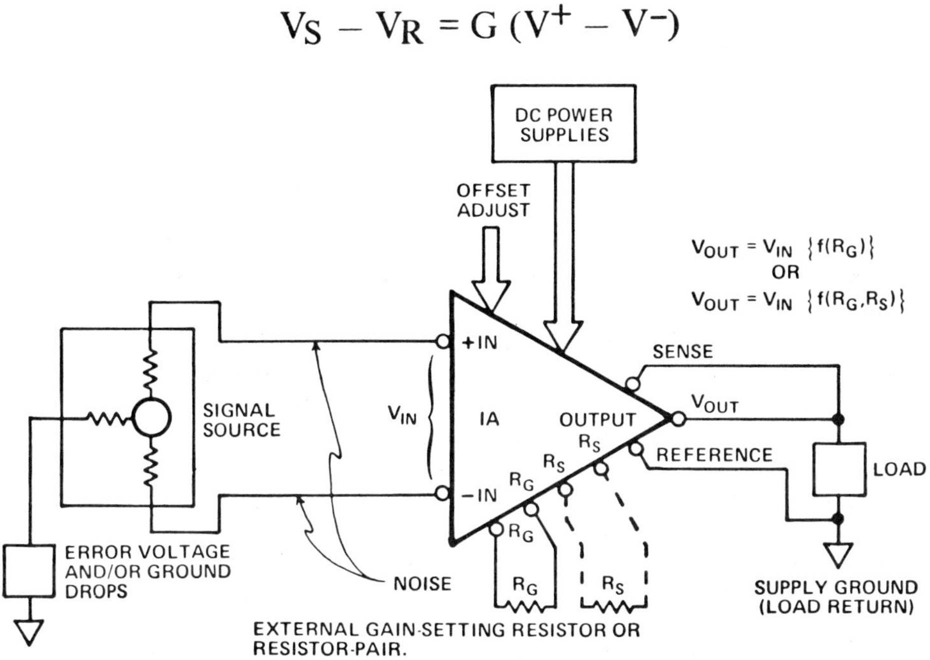


Figure 8-2 Basic Instrumentation Amplifier Block Diagram (Courtesy, Analog Devices)

5. Gain selected by a gain setting resistor $R_G$ and gain trim (stabilizing resistor) $R_S$

**Gain** Gains for instrumentation amplifiers are limited within a range of values. For instance, the gain range for Analog Devices AD522 is from 1 to 1,000. Each device is provided with a gain equation. Gain is set with an external gain resistance. The gain formula for the AD522 is as follows:

$$A_v = 1 + \frac{200{,}000}{R_G} \; V/V$$

where $A_v$ = voltage gain

and $R_G$ = external gain resistance

The calculation for the gain resistance is as follows:

$$R_G = \frac{200{,}000}{A_v - 1}$$

Example for gain $A_v = 100$:

$$R_G = \frac{200{,}000}{100 - 1}$$

$$R_G = 2020\ \Omega$$

**Common-Mode Rejection** (See figure 8-3) Common-mode rejection with instrumentation amplifiers is the measure of output voltage caused when both inputs are changed by the same amount. Common-mode rejection (CMR) is the log of the ratio CMRR. Common-mode rejection ratio (CMRR) is the ratio of signal gain to common-mode gain. In comparison, CMRR of 10,000 is equal to a CMR of 80 dB. Also, a CMRR of 10,000,000 is equal to a CMR of 140 dB. Figure 8-3 shows a simplified CMR. Equal common-mode errors are felt on both inputs. The noninverting input also has a signal present. The instrument amplifier rejects the common-mode errors and only passes the difference signals. CMR usually increases with higher gains. However, at higher gains, bandwidth decreases and CMR becomes dependent on frequency.

**Input Bias and Offset Currents** Input bias currents are used to bias the input transistor stages of the instrument amplifier. FET input devices have lower bias currents. Bias currents increase with temperature by as much as double every 10°C. The difference between the input bias currents is the *offset current*. This difference is the contribution to the offset error.

**Voltage Offset** Voltage offset in instrumentation amplifiers is measured at the output. Offset voltage can be adjusted to zero. Offset voltage is equal to a constant plus a term proportional to gain. The constant is the offset at unity gain.

## THE INSTRUMENTATION AMPLIFIER CIRCUIT (See figure 8-4)

The basic circuit used in instrumentation is the bridge. In the figure, the basic bridge is used along with an instrumentation amplifier. You must

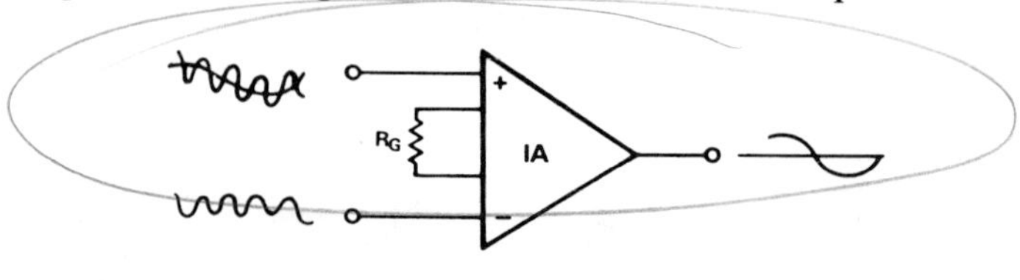


Figure 8-3 Common-Mode Rejection (CMR) (Courtesy, Analog Devices)

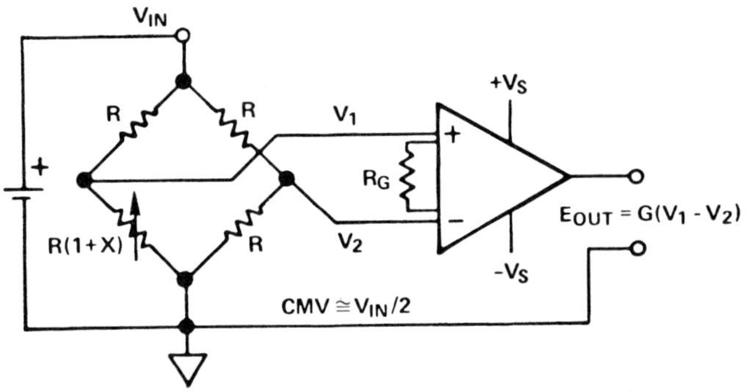


Figure 8-4 Differential Input Instrumentation Amplifier Applied to Bridge Measurement (Courtesy, Analog Devices)

remember, the instrumentation amplifier has a committed gain block. Gain of the amplifier is set by the gain resistor $R_G$ in this configuration. The gain resistor does not have any circuit connections to the inputs.

This device uses a single gain resistor. Some devices use a gain resistor $R_G$ and a gain stabilizing resistor $R_S$ that provides fine adjustment. In some devices, these resistors are both external. Some devices have only one external resistance. Still others contain all the gain resistors within and use jumpers, switches, or digital logic to perform external programming of gain.

In the illustration, when $X$ is zero, the output should be zero. Any error at the output when $X$ is zero is called *common-mode voltage error.* The amplifier should be able to minimize this by its common-mode rejection (CMR).

The amplifier has a balanced differential input. This means that the output will be proportional to the difference between the two input voltages. The input terminals present a high impedance to the input source. Since the amplifier is a differential amplifier, it can be used to amplify transducer output in bridge networks whether the bridge is balanced or unbalanced. Indeed, it can be the heart of every instrumentation signal conditioner.

In many extremely high-precision applications, there is a point at which low levels of drift and common-error voltage are excessive no matter how minute. Even instrumentation amplifiers may prove to be marginal for certain applications. For those situations calling for a very high common-mode rejection ratio (CMRR) along with high voltage, the circuit configuration changes.

In figure 8-5, the single amplifier circuit is reconfigured with a split dc power supply. The circuit will reduce dc common-mode voltage. It will not reduce drift or other common-mode variations.

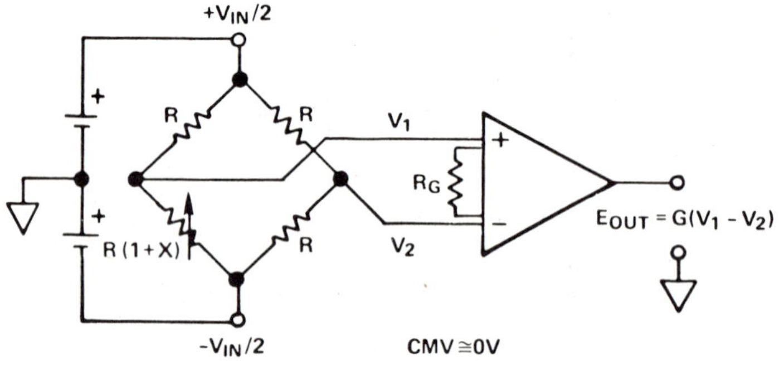


Figure 8-5 Center Tapped Power Supply Used to Minimize Common-Mode Voltage (Courtesy, Analog Devices)

## GROUNDING (See figure 8-6)

Instrumentation amplifiers all have differential inputs. There are, however, no ground paths for bias current. These must be provided or the bias current may charge stray capacitance and in turn cause the amplifier output to saturate or to drift. When inputs such as transformers, transducers, or simply ac coupled sources are being amplified, a ground must be supplied. In the figure, grounds are placed on the input devices and made common with load ground (power supply).

## INSTRUMENTATION AMPLIFIER (IA) MODULE FABRICATION

While the operational amplifier can be used as an instrumentation amplifier, it has its drawbacks. With the use of a single op amp differential stage, there are unbalanced and low input impedances and low common-mode rejection (CMR). Another instrumentation amplifier, the two op amp circuit, provides high input impedance but gain range is neglected. The most common of the configurations is the classic three op amp approach. It provides the engineer with the gain desired without increasing common-mode error. The circuit in figure 8-7 illustrates the three op amp instrumentation amplifiers. This circuit is generally used when the instrumentation amplifier (IA) module is not available. This circuit is constructed with discrete components and three op amps. The instrumentation amplifier (IA) module usually takes the three op amp approach under two methods of fabrication, hybrid and monolithic.

**The Hybrid Method** (See figure 8-8) The three op amp approach to hybrid fabrication is used to create the Analog Devices AD522, an

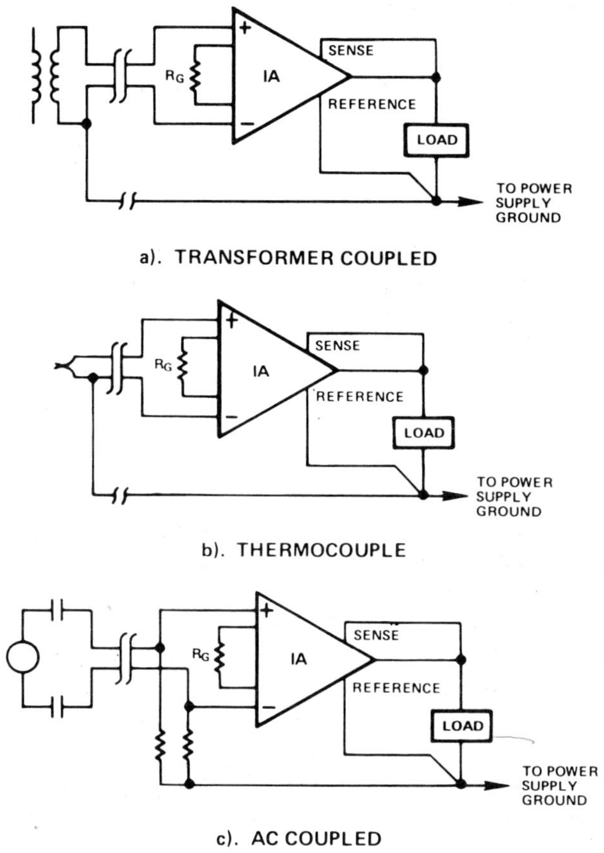


Figure 8-6 Ground Returns for Bias Current (Courtesy, Analog Devices)

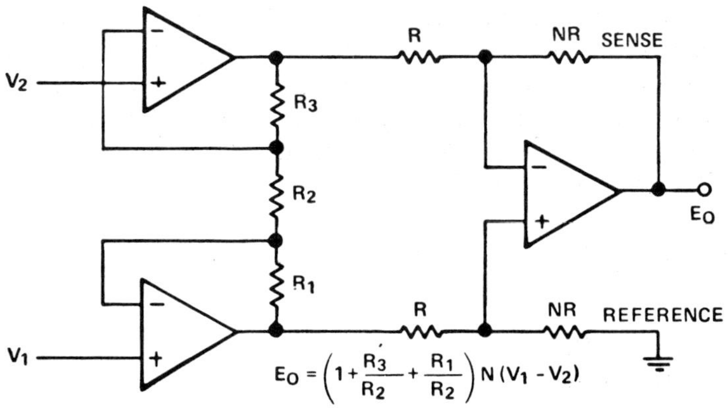


Figure 8-7 Three Op Amp Instrumentation Amplifier Using Discrete Components (Courtesy, Analog Devices)

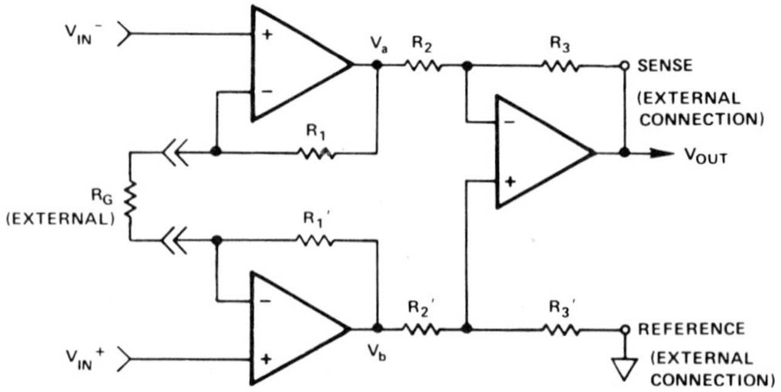


Figure 8-8 The Three-Amplifier Instrumentation Configuration as a Hybrid Device (Courtesy, Analog Devices)

(IA) module. The input stage is a differential input/output gain stage, while the third op amp is a subtractor. Gain is set by adding an external gain resistor $R_G$. If the sense terminal is connected to $V_{OUT}$ and the reference terminal is connected to power ground, the output voltage will appear between the two terminals. Gain may be taken without regard to increasing the common-mode error signal. Common-mode rejection (CMR) will increase in direct proportion to ghe gain. Common-mode signals are only amplified by a factor of 1 regardless of gain. No common-mode current flows in the amplifier. (Remember that op amps have no significant difference of potential across the inverting and noninverting inputs.)

The circuit fabricated by the hybrid method is similar (not quite identical) to the discrete three op amp circuit shown in figure 8-7.

**The Monolithic Method (See figure 8-9)** Typical of the monolithic instrumentation amplifier (IA) modules is the Analog Devices AD521. The technique is based on minimum active device count. A differential voltage $V_{IN}$ appears across the gain resistor $R_G$. The imbalance flows through stabilizing resistor $R_S$. The output differential voltage is equal to $\Delta I \times R_S$. Therefore

$$\frac{V_{OUT}}{R_S} = \frac{V_{IN}}{R_G}$$

or

$$\frac{V_{OUT}}{V_{IN}} = \frac{R_S}{R_G}$$

also

$$\Delta I = \frac{\Delta V_{OUT}}{R_S}$$

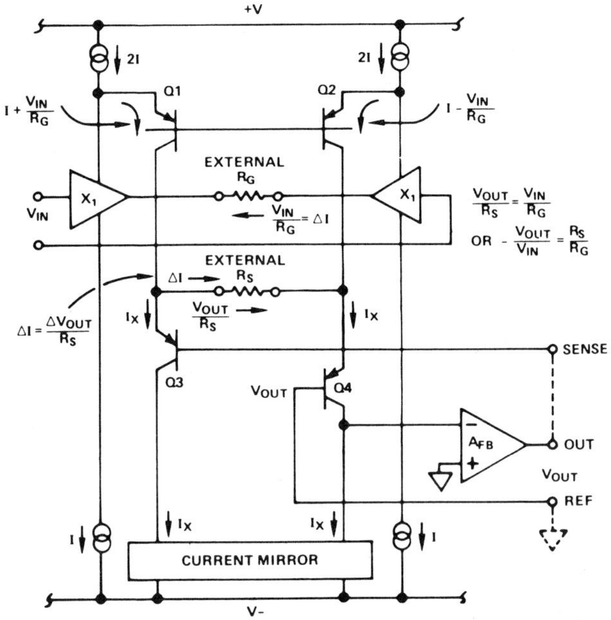


Figure 8-9 Simplified Monolithic Instrumentation Amplifier (Courtesy, Analog Devices)

$$\Delta I = \frac{V_{IN}}{R_G}$$

where $\Delta I$ = changes in current through input transistors Q1 and Q2

## THE ISOLATION AMPLIFIER

An isolation amplifier is a special-purpose instrumentation amplifier used for applications that have extremely high voltage excitation or bridge voltages (to thousands of volts). The isolation amplifier's input is galvanically isolated from the power supply and the output.

Isolation is required when low signals are to be measured in the presence of high voltages. Isolation is also required to eliminate measurement errors such as high current transients. A further use of the isolation amplifier is avoid ground loops and pickup problems in hostile or critical environments such as hospital rooms. The isolation amplifier is primarily used in applications that require ohmic or galvanic isolation of input and output circuitry.

**Function AND Operation (See figure 8-10)** Typical of the isolation amplifiers or isolators is the Analog Devices Model 289. This device has a floating input that is isolated from the power terminals and the output.

Power for the device is plus and minus dc to an isolated internal power supply. An internal oscillator converts the dc to ac. It is further coupled by a transformer to a shielded input stage, where it is converted back to dc for that stage. The ac carrier is modulated by the amplifier output transformer coupled to a demodulator, filtered, and amplified. Power for the output stage is also isolated.

The first-stage input amplifier functions as an instrumentation amplifier. It is protected by resistors from overloads and sensitive sources from supply voltage.

In figure 8-10, the amplifier is shown separated in three stages: input, output, and power. In Model 289 these three sections (or ports, as they are called) are isolated to provide the galvanic and ohmic isolation required. The isolators have high common-mode rejection (CMR), low drift, and can handle extremely high common-mode voltages (CMV).

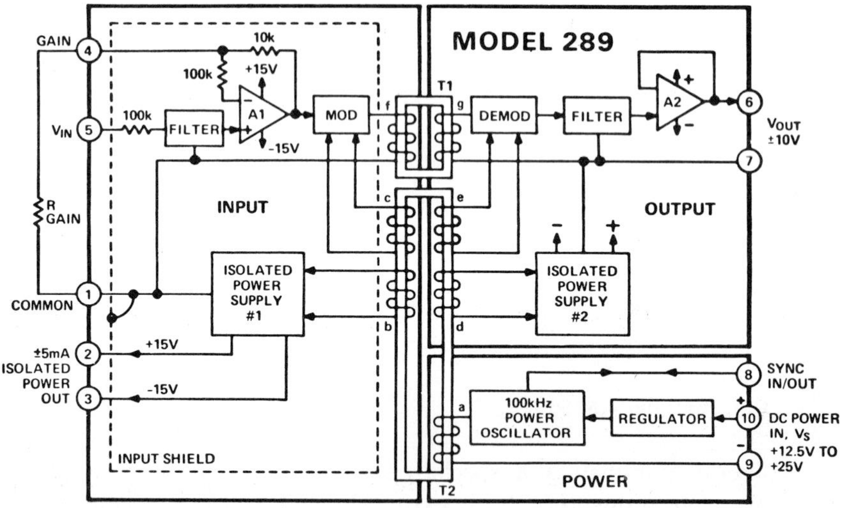


Figure 8-10 An Isolation Amplifier (Courtesy, Analog Devices)

In figure 8-11, an application is presented. A high voltage excites a bridge network. The bridge feeds the signal to the isolator input stage. DC power is converted to ac and coupled across the isolated section, where it is rectified to provide power for the amplifier stage. signals modulate the carrier and are coupled to the output stage, modulated, and placed at the output.

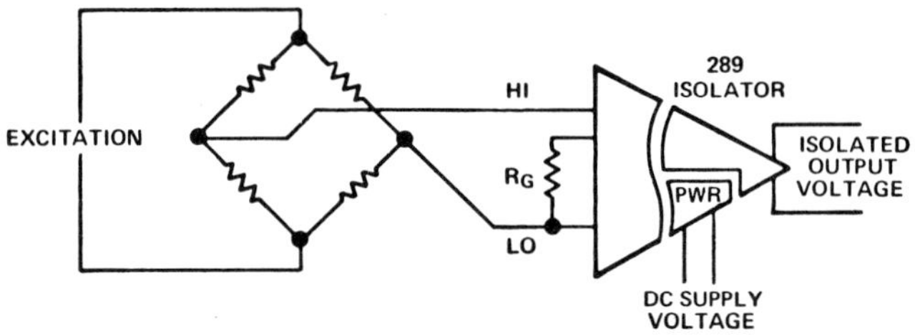


**Figure 8-11** Application of an Isolation Amplifier (Courtesy, Analog Devices)

# 9
# Special Design Applications

It would take a dozen books of this size to describe every op amp and every application. Even then, the number of applications would increase the day after publication. The author does not intend that this book compete in special applications. There are many extremely fine application books on the market. Indeed, the best and most complete of these are usually written by the manufacturer of the particular device.

The purpose of this chapter is to provide typical circuits, from which the reader may choose. The chapter does not replace device applications handbooks but may fill some void between them and standard textbooks. Applications in this chapter are separated into types such as amplifiers.

Circuit types presented in this chapter are in alphabetical order as follows:

Amplifiers
Detectors
Differentiators/Integrators
Filters
Generators

Instrument Transmitters
Logic and Related Circuits

Mixers
Oscillators

Power Monitors
Regulators
Sample and Hold Circuits
Tachometers

Schematics in this chapter are reprinted from National Semiconductor Corporation's *Linear Data Book 1M-RRD85M31* with permission of the Director of Advertising.

## AMPLIFIERS

The schematics following illustrate a series of amplifiers. Circuits presented are power, ac, dc, transconductance, buffer, bridge, photo cell, squaring, photo diode, and hi-$Z$ amplifiers.

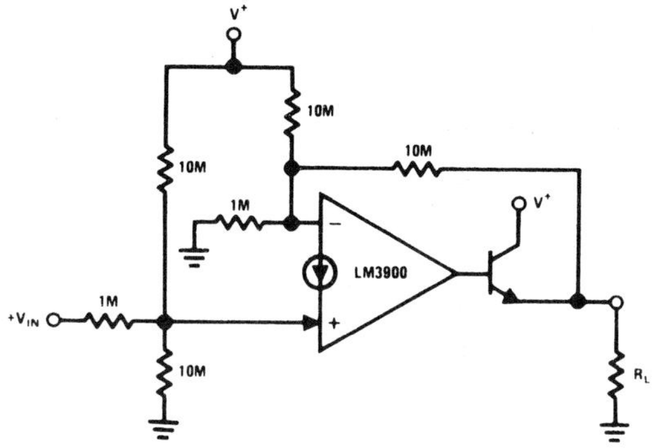


Power Amplifier

Power Amplifier

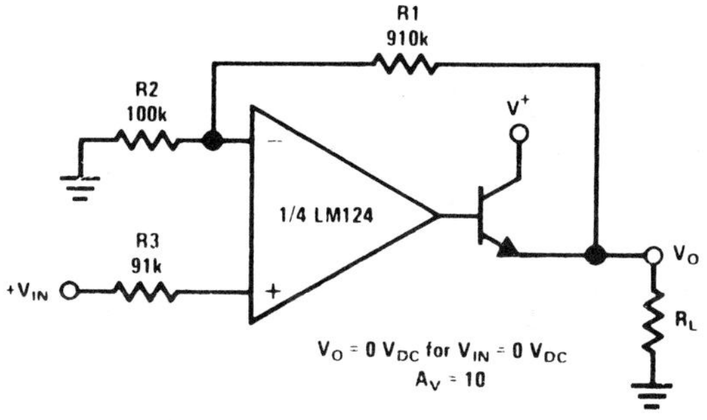


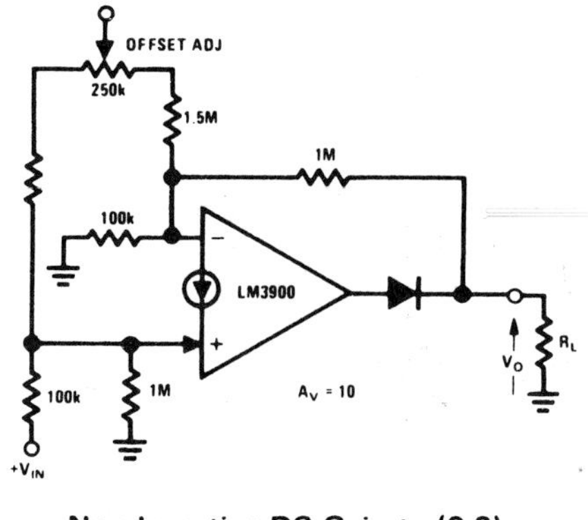


Non-Inverting DC Gain to (0,0)

### AC Coupled Non-Inverting Amplifier

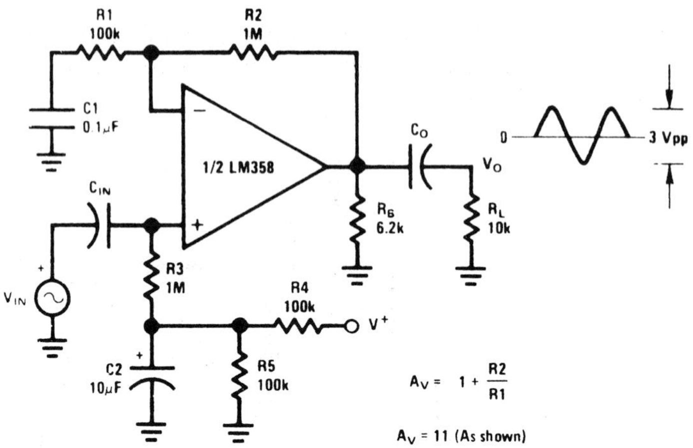


$A_V = 11$ (As shown)

### DC Summing Amplifier
### ($V_{IN'S} \geq 0\ V_{DC}$ AND $V_O \geq 0\ V_{DC}$)

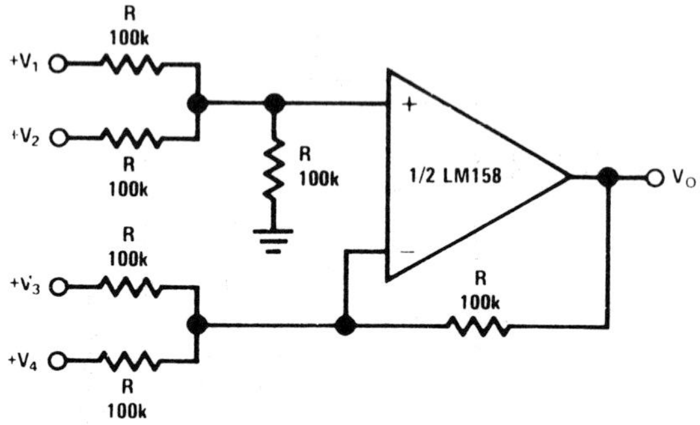


Where: $V_O = V_1 + V_2 - V_3\ \ V_4$
$(V_1 + V_2) \geq (V_3 + V_4)$ to keep $V_O > 0\ V_{DC}$

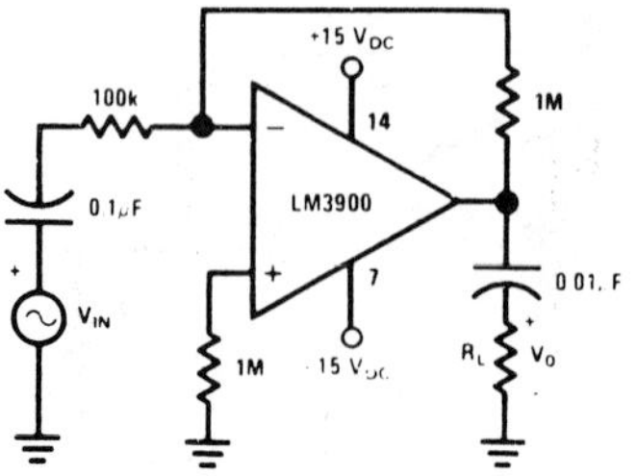


**AC Amplifier**

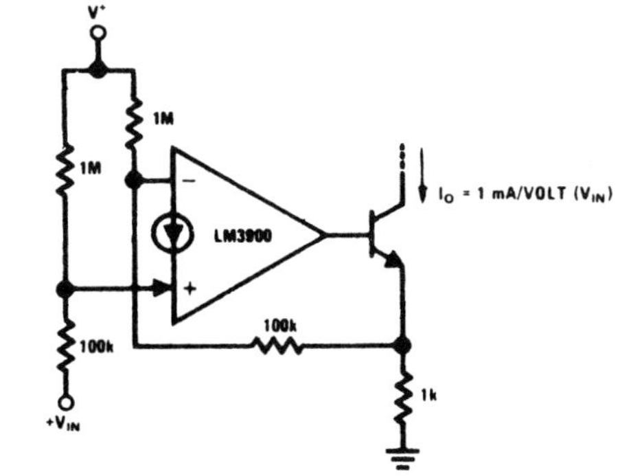


**Voltage-Controlled Current Sink (Transconductance Amplifier)**

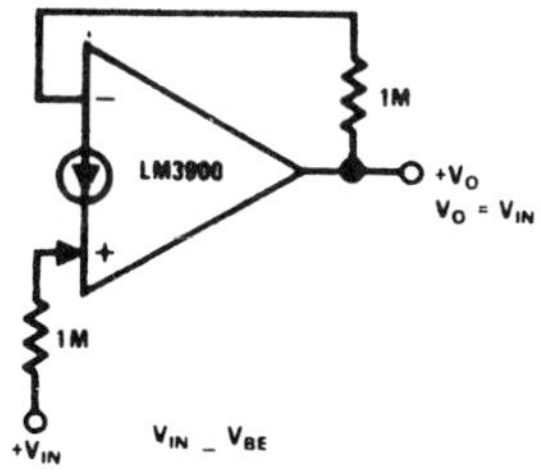


**Buffer Amplifier**

Bridge Amplifier

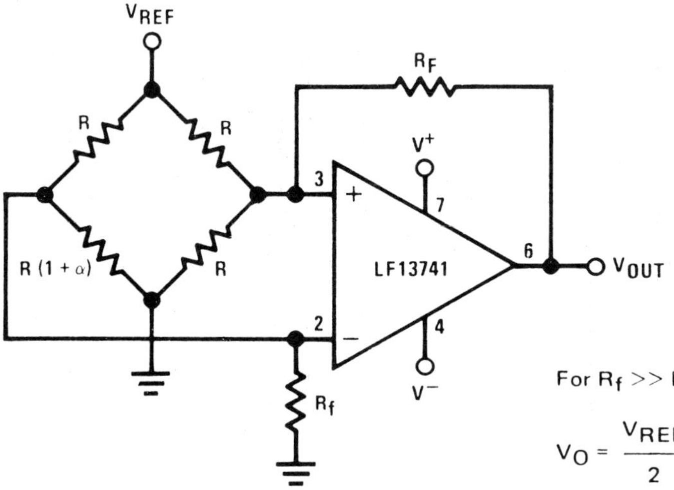


For $R_f >> R$

$$V_O = \frac{V_{REF}}{2}\left(\frac{R_f}{R}\right)\left(\frac{\alpha}{1+\alpha}\right)$$

**Photo Voltaic-Cell Amplifier**

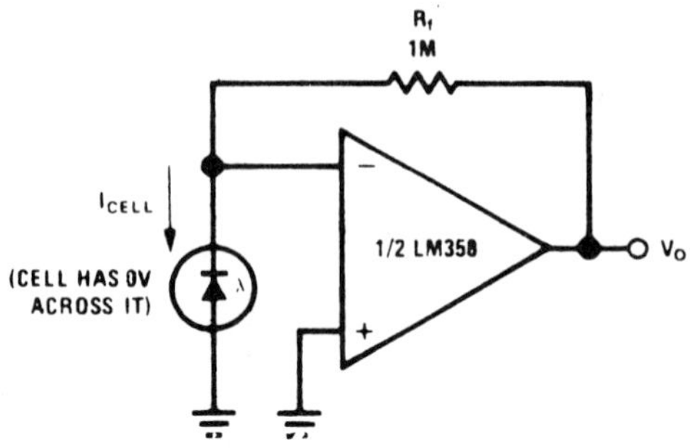


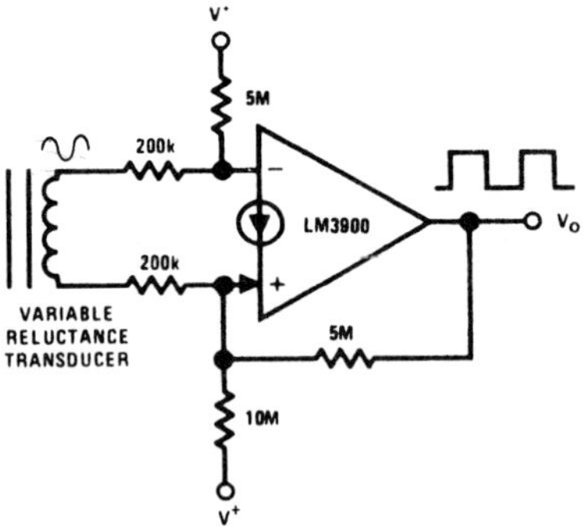


**Squaring Amplifier (W/Hysteresis)**

**Photo Diode Amplifier**

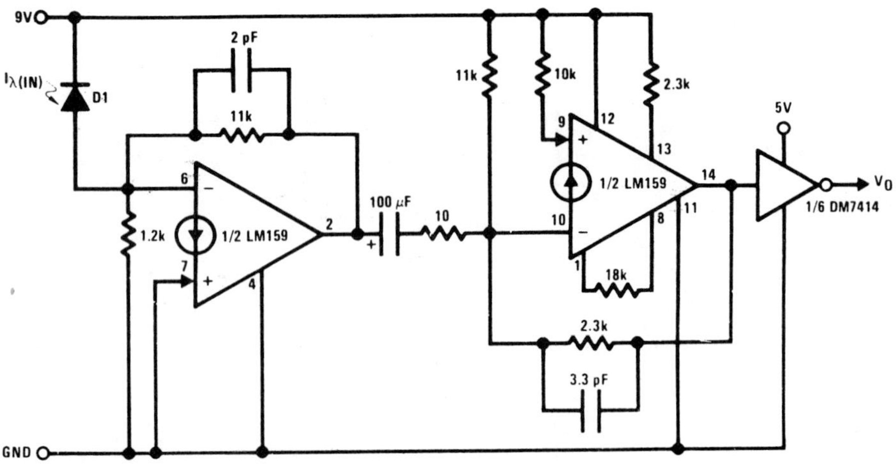


D1~RCA N-Type Silicon P-I-N Photodiode

**High Input Z Adjustable-Gain DC Instrumentation Amplifier**

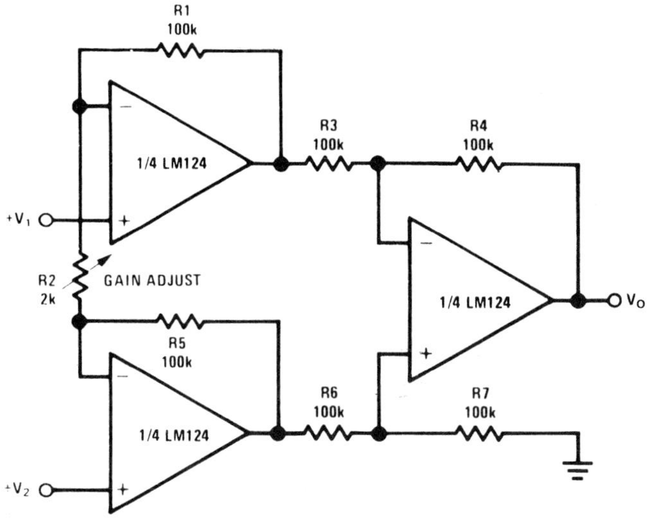


If R1 = R5 & R3 = R4 = R6 = R7 (CMRR depends on match)

$$V_O = 1 + \frac{2R1}{R2} (V_2 \quad V_1)$$

As shown $V_O = 101 (V_2 \quad V_1)$

## DETECTORS

The schematics following are a series of detector circuits. Circuits presented are low and ultra-low draft-peak detectors.

**Ultra-Low Drift Peak Detector**

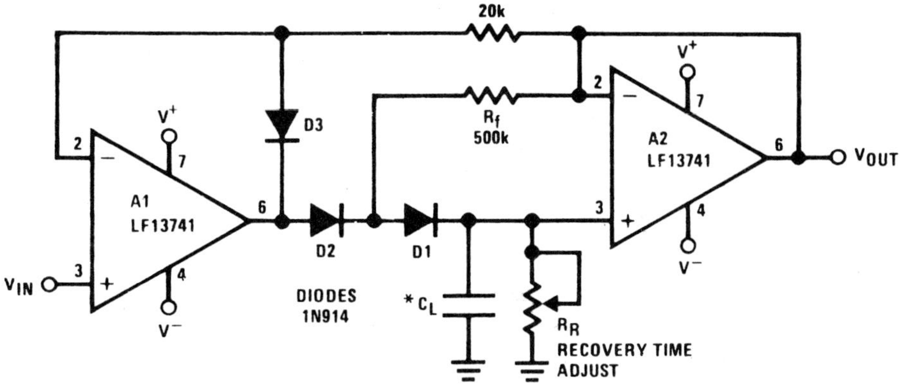

Low Drift Peak Detector

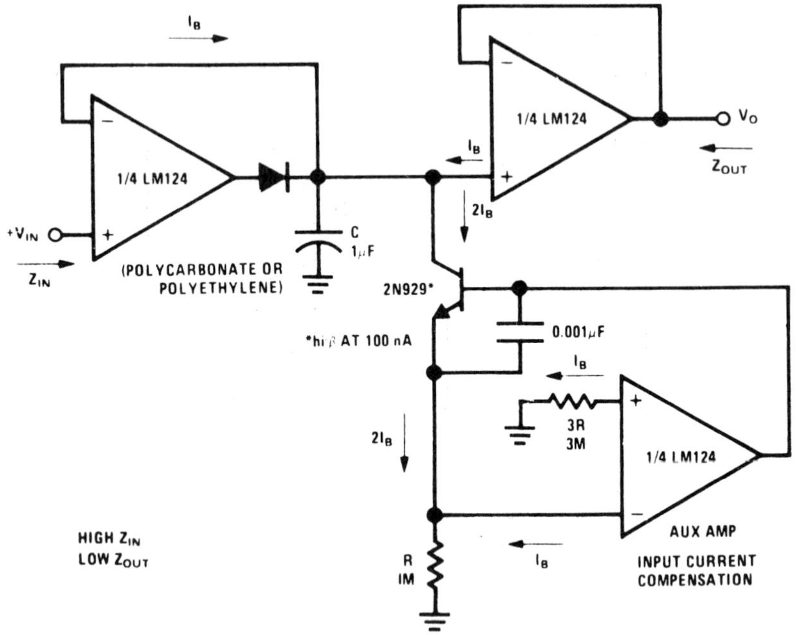


Low Drift Peak Detector

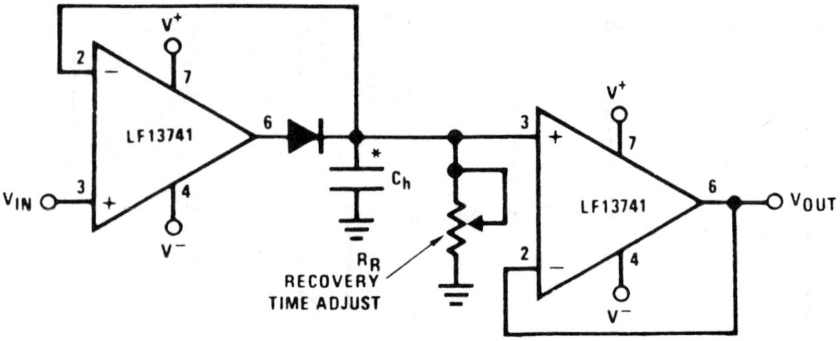


*Low leakage capacitor

# DIFFERENTIATORS/INTEGRATORS

The schematics following are a series of differentiators and integrators. Circuits presented are the differentiator, the difference integrator, and a long-time integrator.

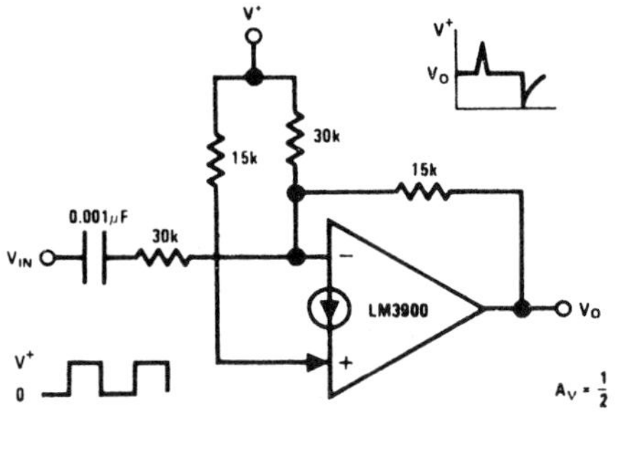


**Differentiator (Common-Mode Biasing Keeps Input at $+V_{BE}$)**

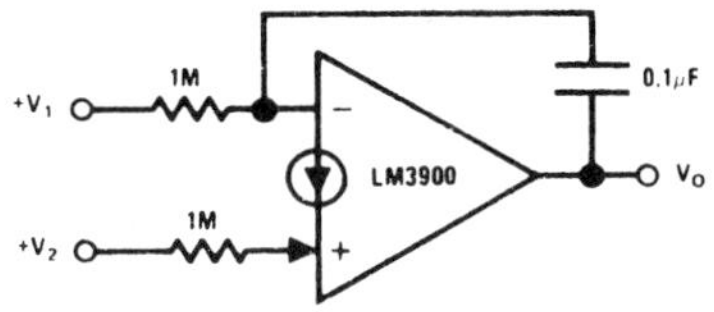


**Difference Integrator**

Long Time Integrator

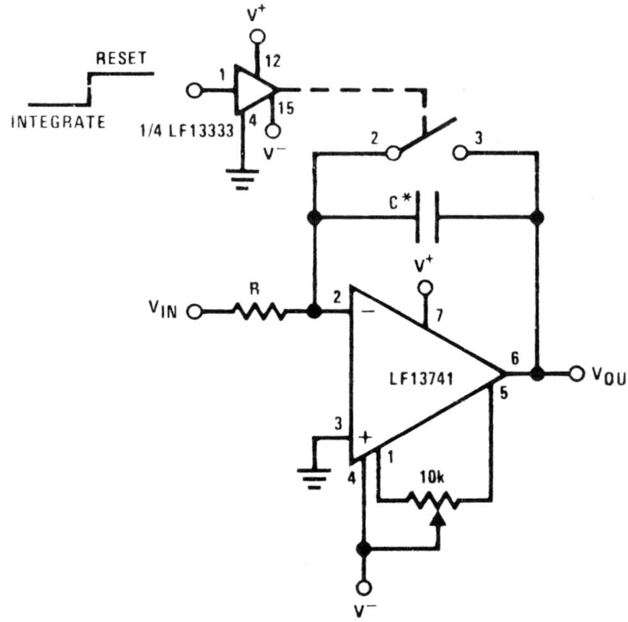


$V_{OUT} = \frac{1}{RC} \int_{t_1}^{t_2} V_{IN} \, dt$

*Low leakage capacitor

## FILTER CIRCUITS

The schematics following are a series of filters. Circuits presented are single, dual, and three-amplifier bandpass filters.

**Bandpass Active Filter**

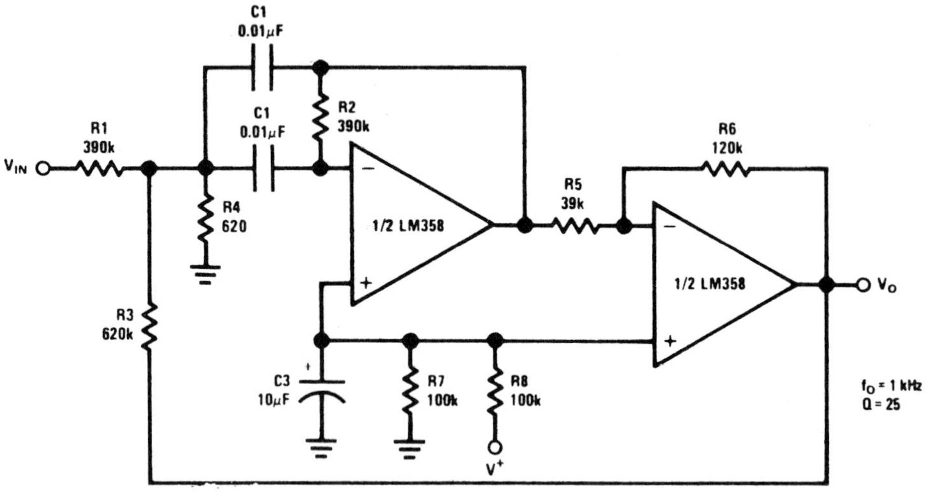


**DC Coupled Low-Pass RC Active Filter**

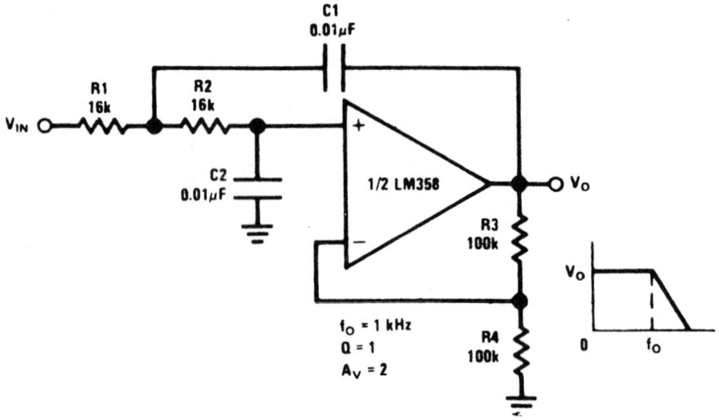

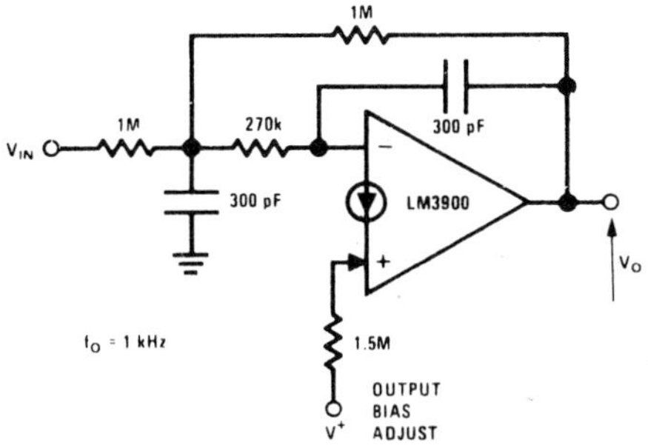


Low Pass Active Filter

## High Performance 2 Amplifier Biquad Filter(s)

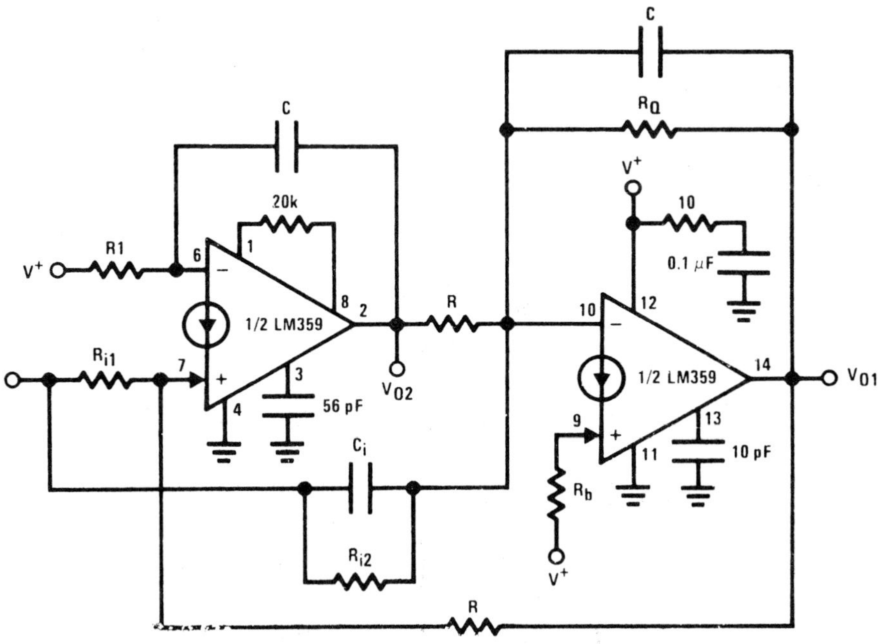

**"BI-QUAD" RC Active Bandpass Filter**

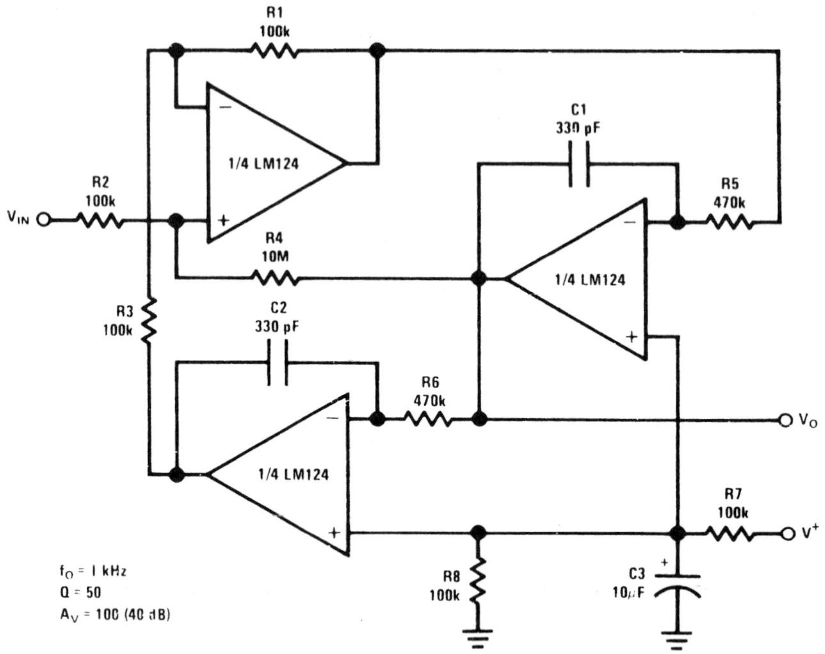


## GENERATORS

The schematics following are a series of generators. Circuits presented are pulse, squarewave, sawtooth, triangular, and staircase.

**Pulse Generator**

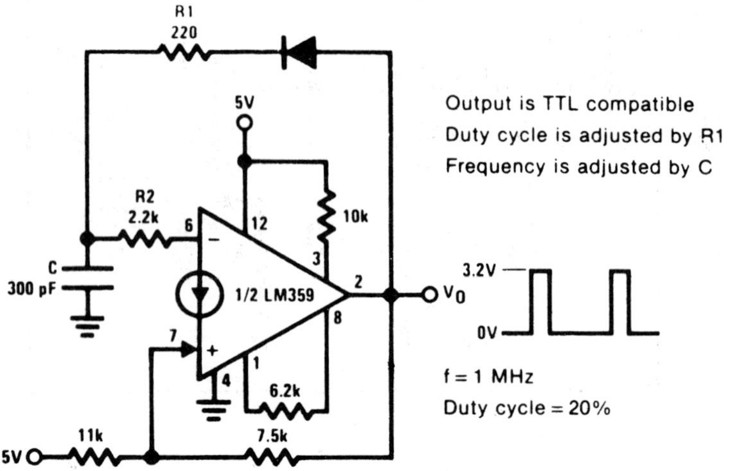

## Squarewave Generator

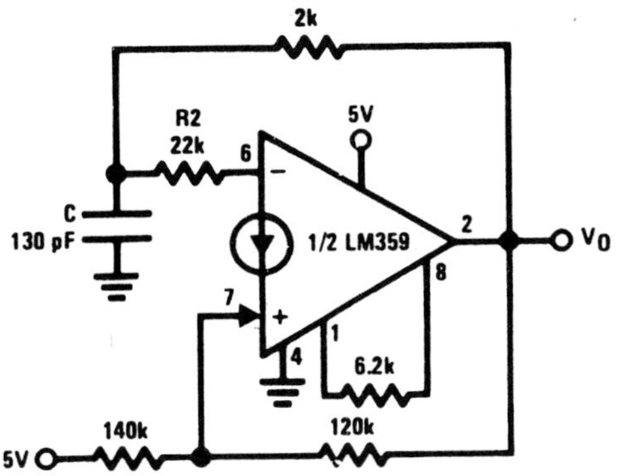


f = 1 MHz

Output is TTL compatible

Frequency is adjusted by R1 & C (R1<<R2)

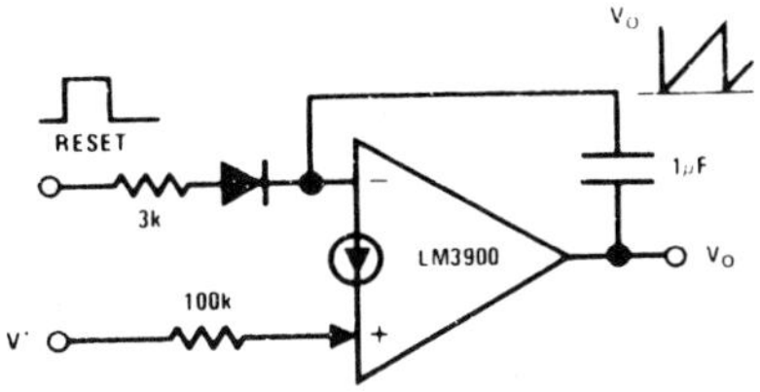


**Sawtooth Generator**

## Triangle Waveform Generator

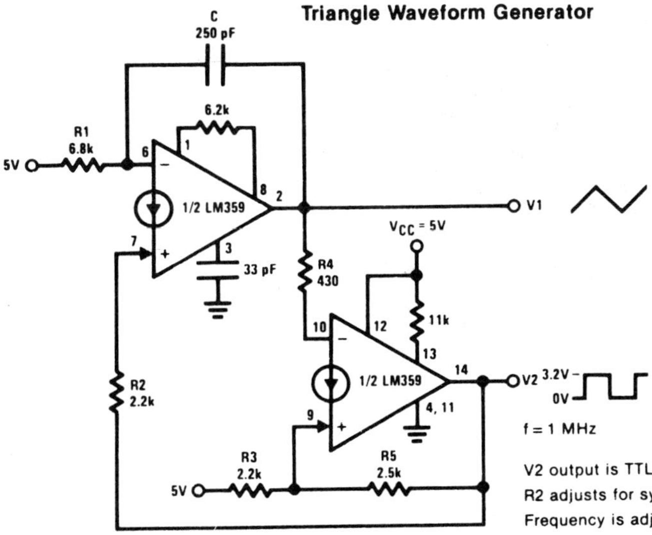


f = 1 MHz

V2 output is TTL compatible

R2 adjusts for symmetry of the triangle waveform

Frequency is adjusted with R5 and C

**Pulse Generator**

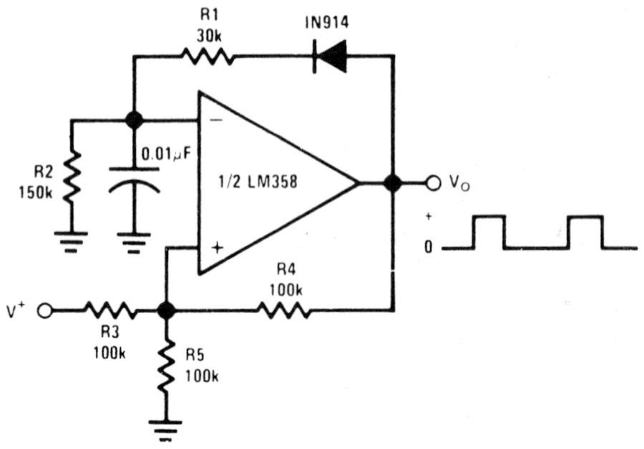


**Pulse Generator**

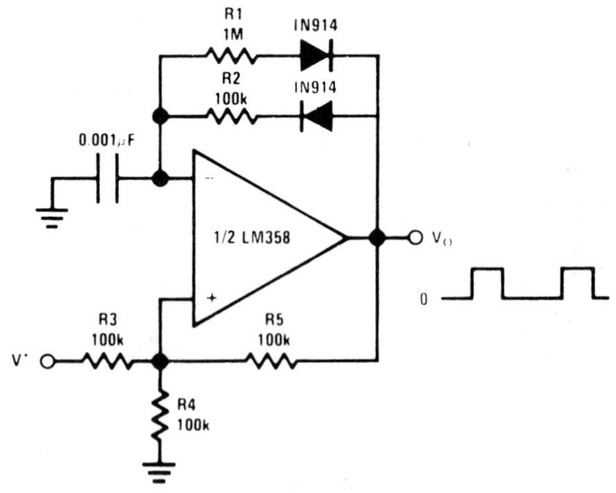

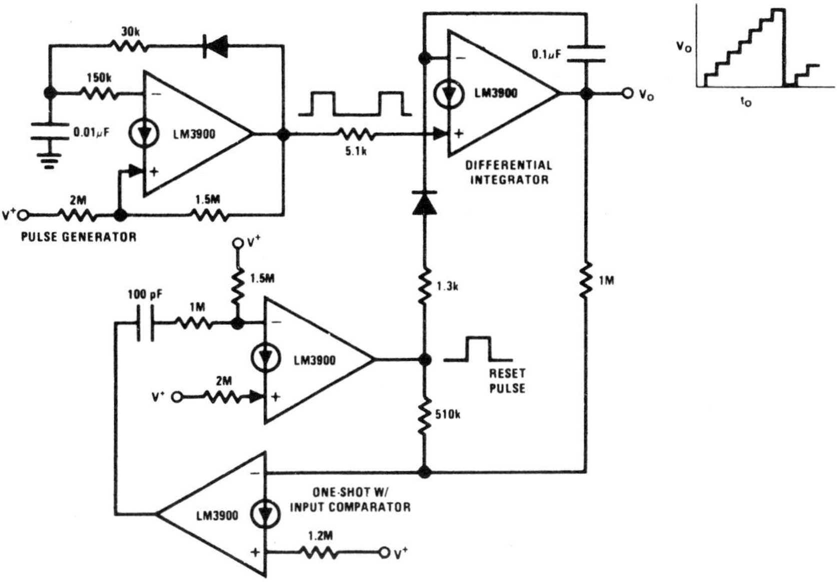


**Free-Running Staircase Generator/Pulse Counter**

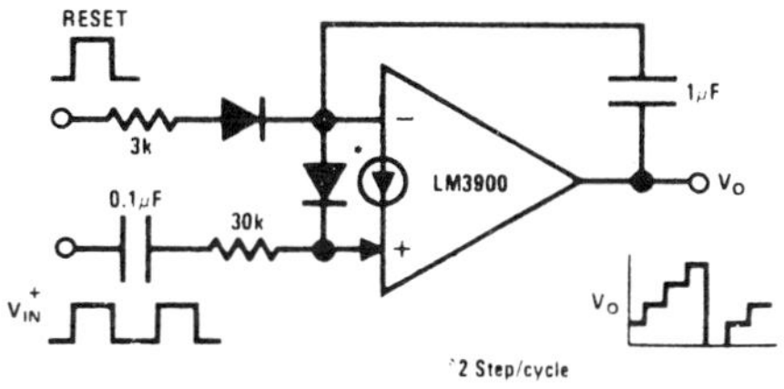


**Staircase Generator**

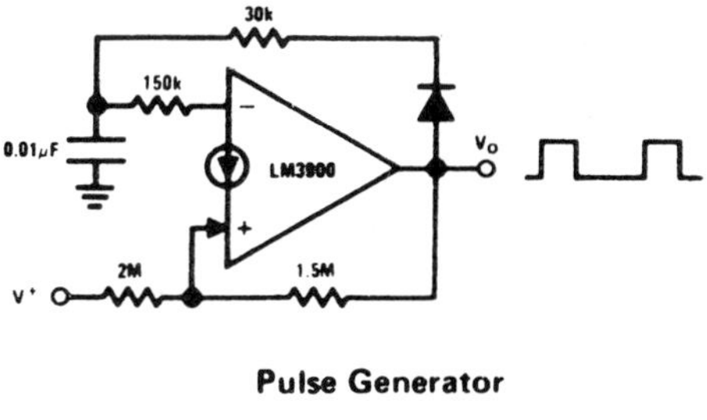


**Pulse Generator**

## INSTRUMENT TRANSMITTERS

The schematics following are a series of instrument transmitters. Circuits presented are a precision thermocouple transmitter and a transmitter for bridge sensor.

**precision thermocouple transmitter**

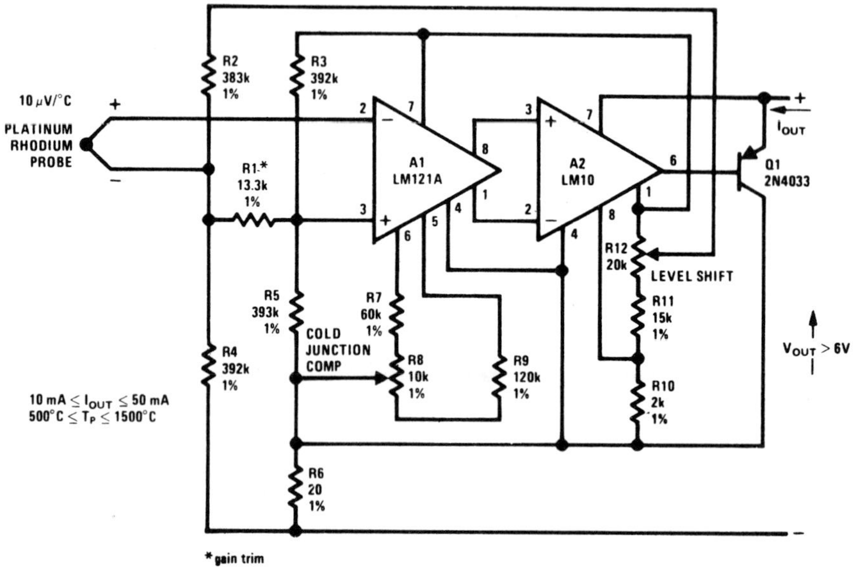


**transmitter for bridge sensor**

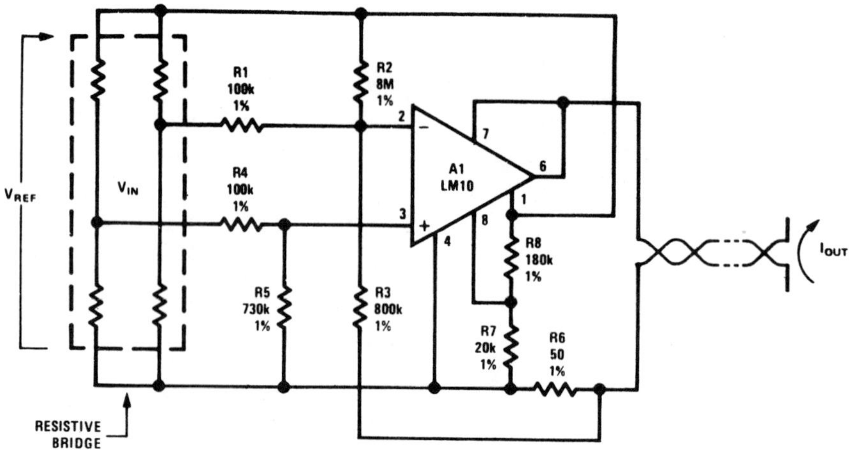

# LOGIC AND RELATED CIRCUITS

The schematics following are related to logic. Circuits presented are gates, multivibrators, triggers, comparators, and lamp drivers.

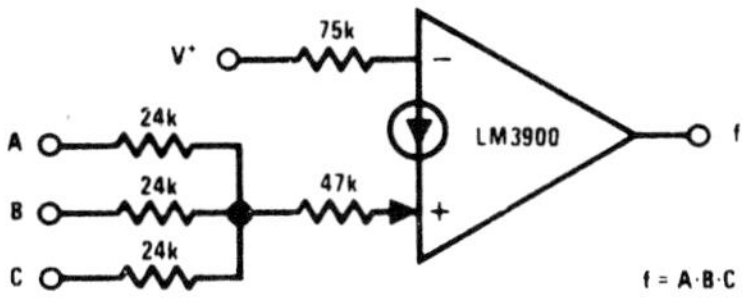


"AND" Gate

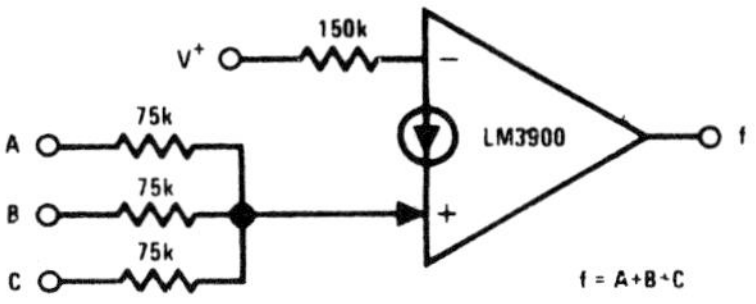


"OR" Gate

Driving TTL

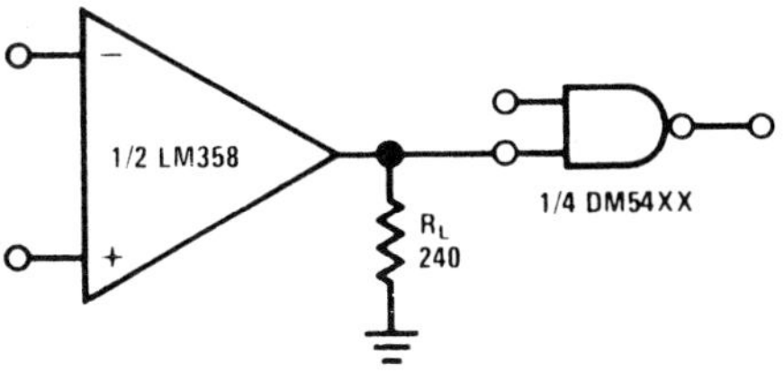


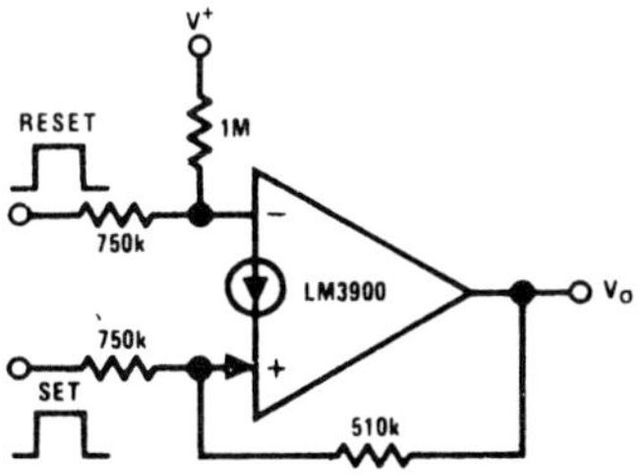


Bi-Stable Multivibrator

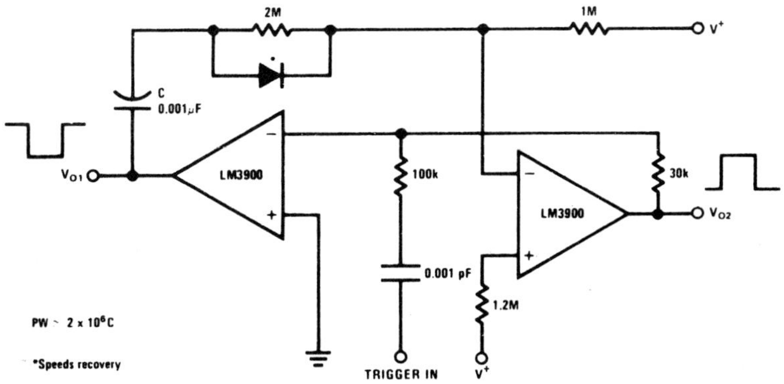


One-Shot Multivibrator

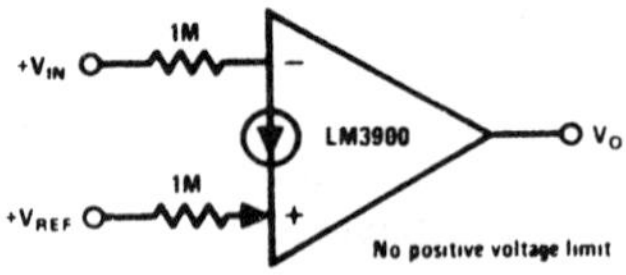


Comparator

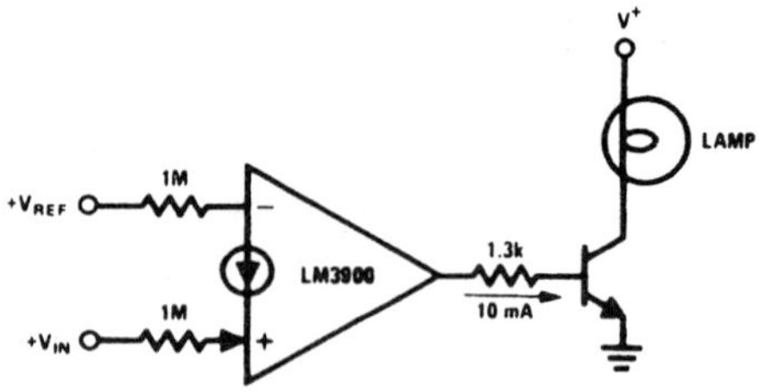


Power Comparator

Comparator with Offset Adjust for Hi-Z Inputs

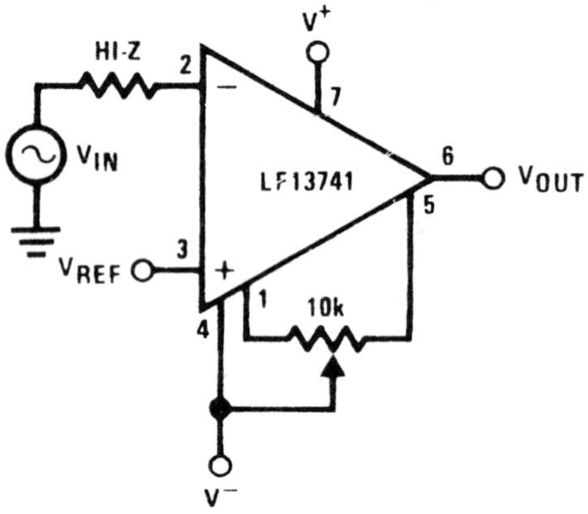


$$V^- + 3V \leq V_{IN} \leq V^+ + 0.1V$$

Comparator with Hysteresis

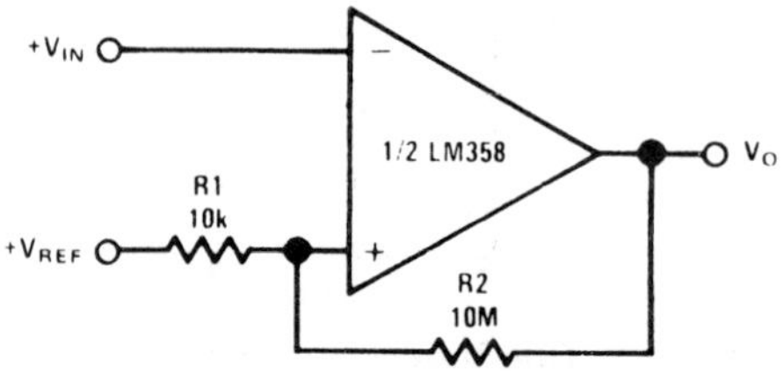

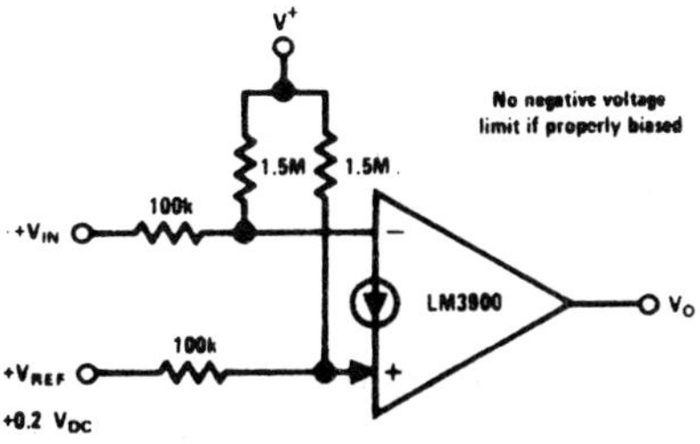


**Low-Voltage Comparator**

Lamp Driver

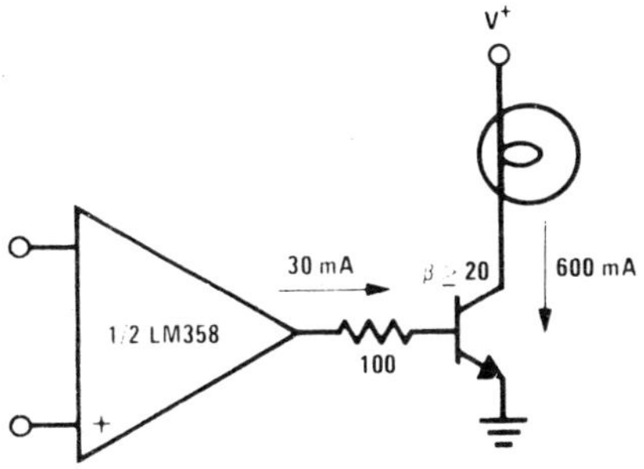


LED Driver

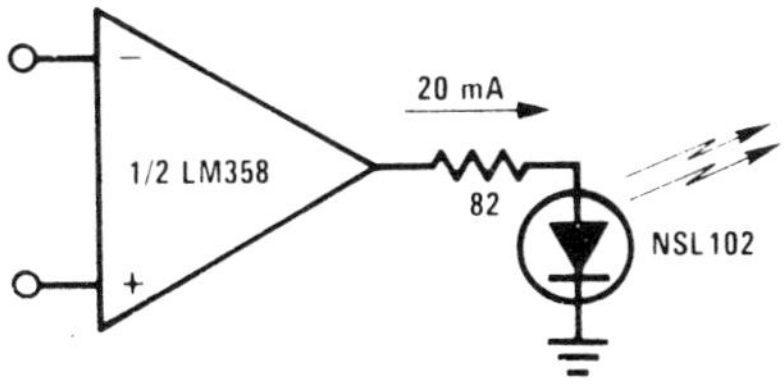


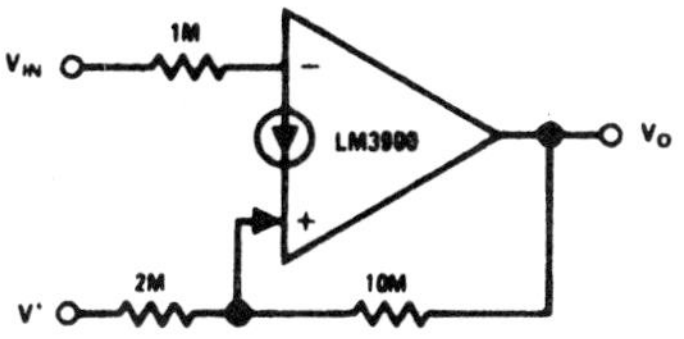


**Schmitt-Trigger**

## MIXERS

The schematics following are a series of mixer circuits. Circuits presented are a dc control and a low-frequency mixer.

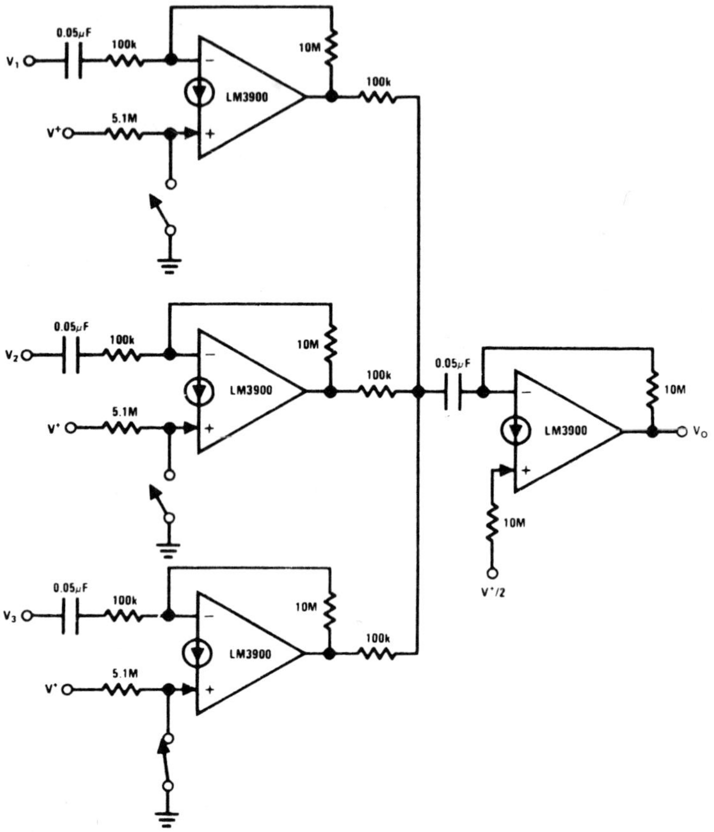


**Channel Selection by DC Control (or Audio Mixer)**

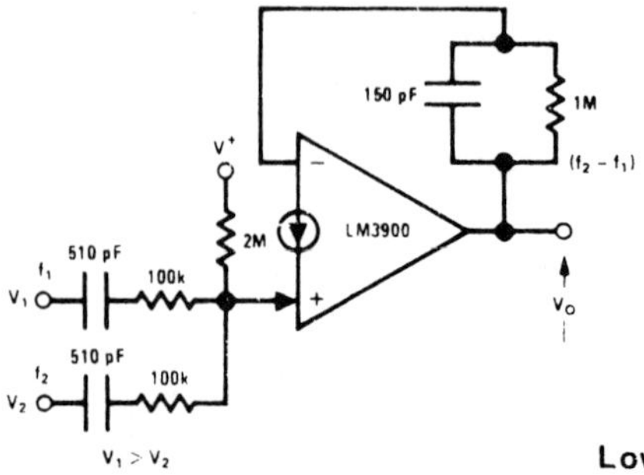


**Low-Frequency Mixer**

# OSCILLATORS

The schematics following are a series of oscillators. Circuits presented are voltage-controlled, crystal-controlled, and squarewave.

Voltage Controlled Oscillator (VCO)

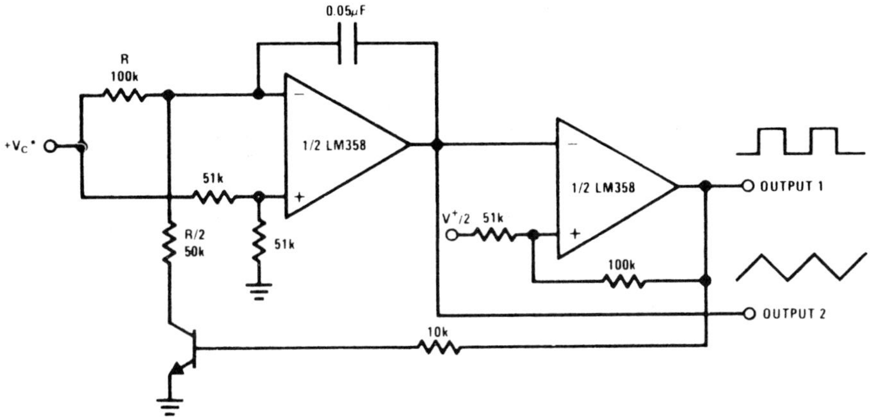


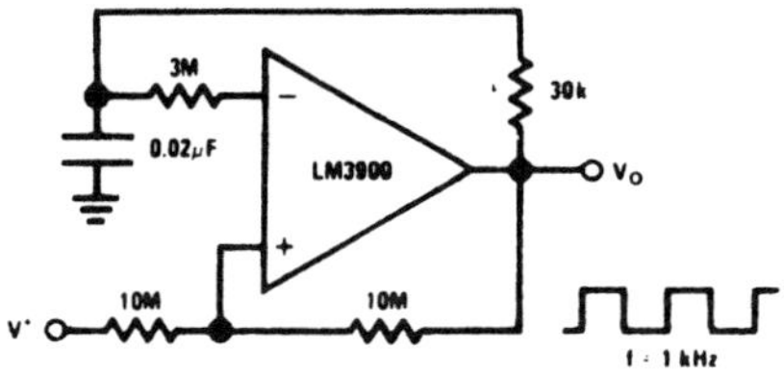


Square-Wave Oscillator

Squarewave Oscillator

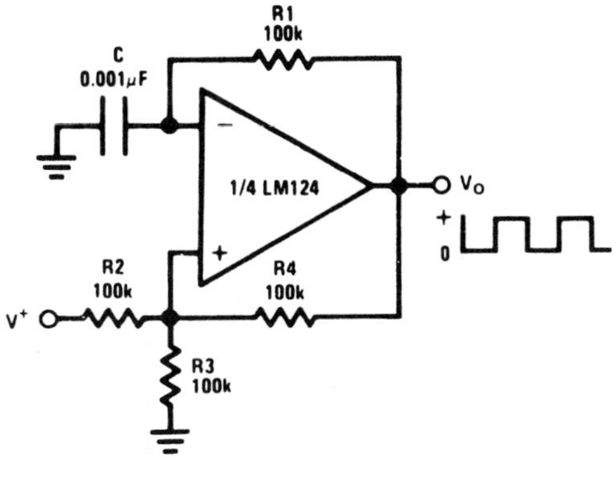

**Crystal Controlled Sinewave Oscillator**

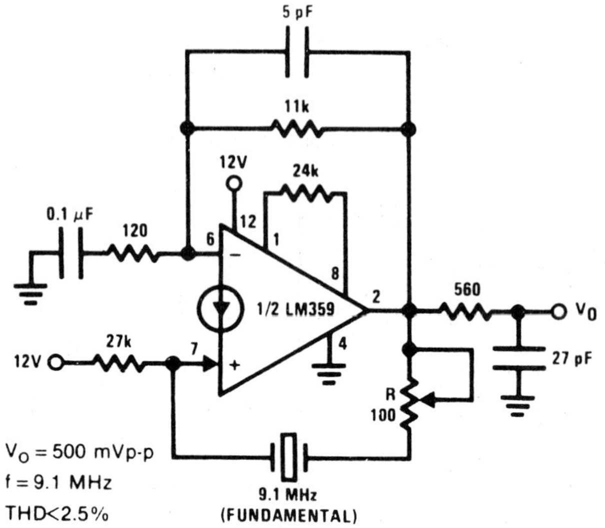


## POWER MONITORS

The schematics following are a series of power monitors. Circuits presented are voltage and current monitors, and battery threshold/level indication.

**double-ended voltage monitor**

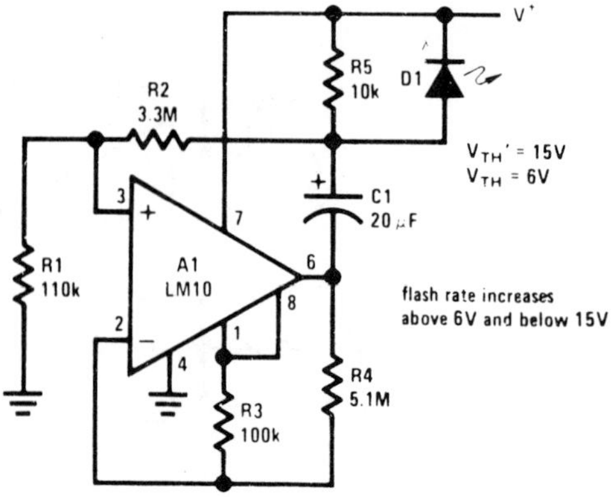

## single-cell voltage monitor

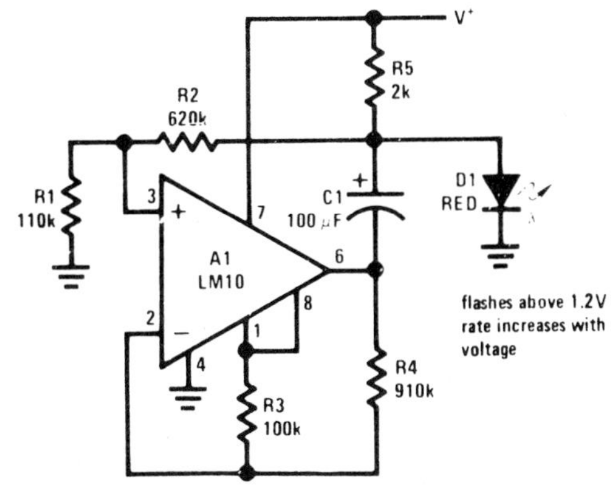


## isolated voltage sensor

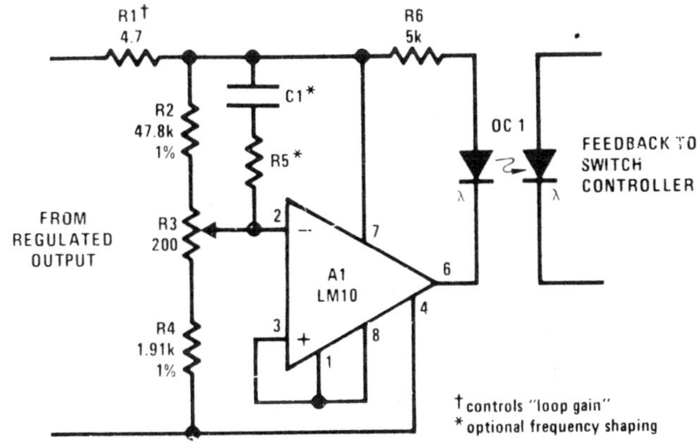


## battery-threshold indicator

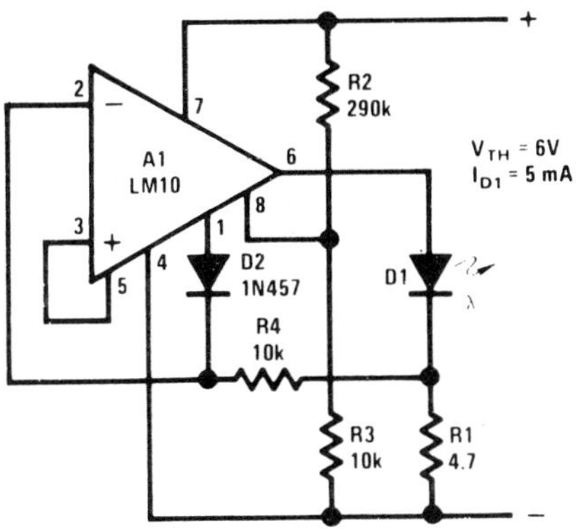

**battery-level indicator**

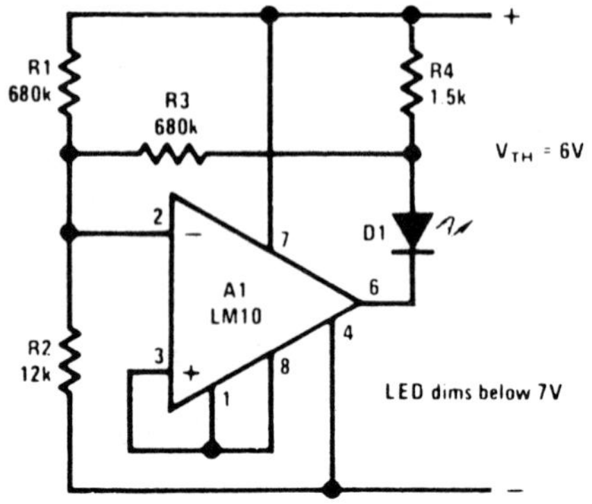


**Supply Current Monitor**

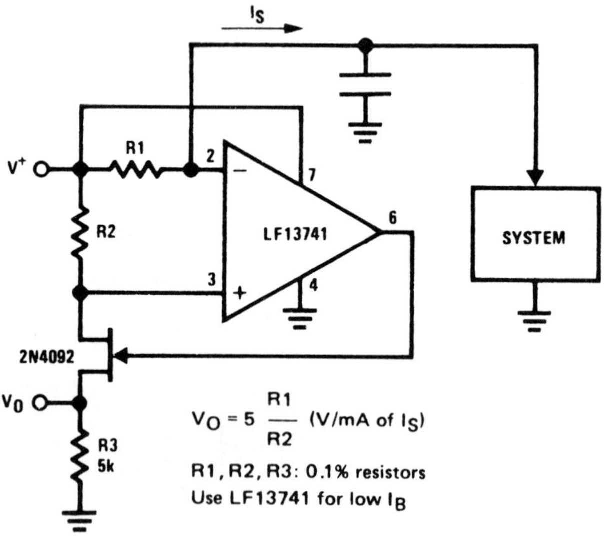


$$V_O = 5\ \frac{R1}{R2}\ \text{(V/mA of } I_S\text{)}$$

R1, R2, R3: 0.1% resistors

Use LF13741 for low $I_B$

# REGULATORS

The schematics following are a series of regulators. Circuits presented are Hi $V_{IN}$ self-regulator and voltage and current regulators.

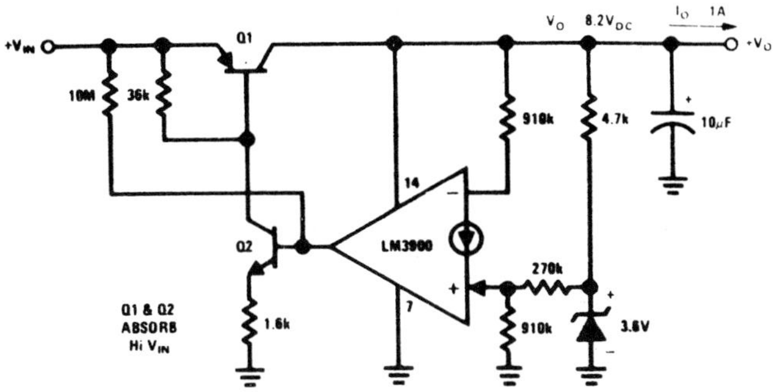


Hi $V_{IN}$, Lo ($V_{IN} - V_O$) Self-Regulator

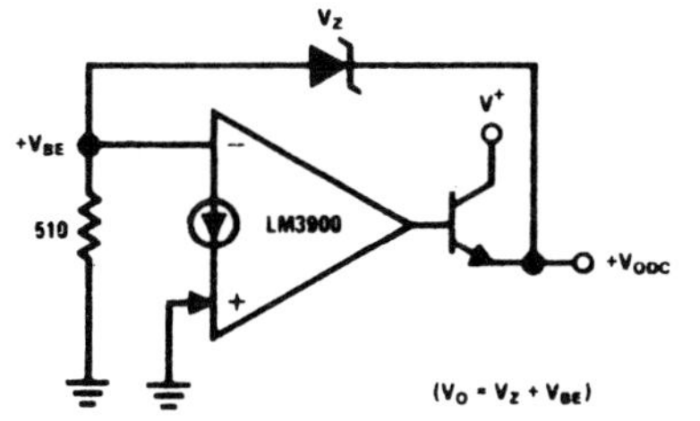


Voltage Regulator

shunt regulator

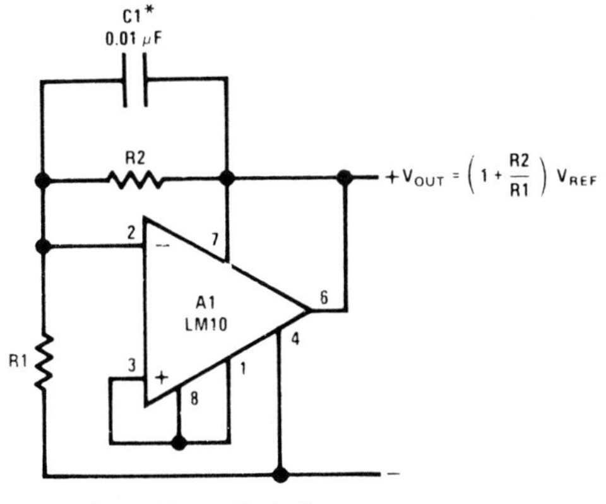


* required for capacitive loading

**current regulator**

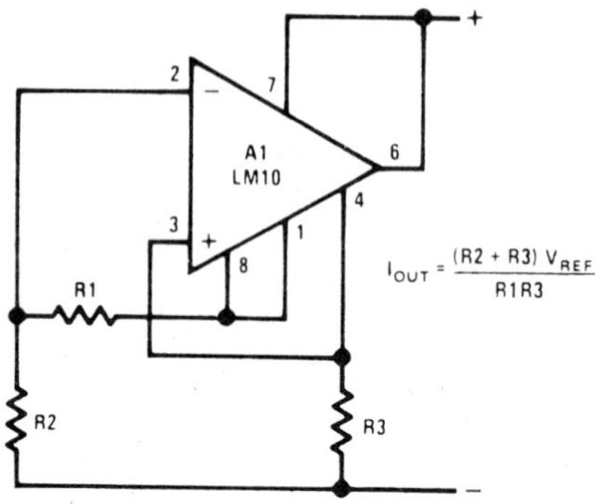


## SAMPLE AND HOLD CIRCUITS

The schematics following are a series of sample and hold circuits. The circuits presented are fast sample and hold, high accuracy sample and hold, and low drift ramp and hold circuits.

**Fast Sample and Hold**

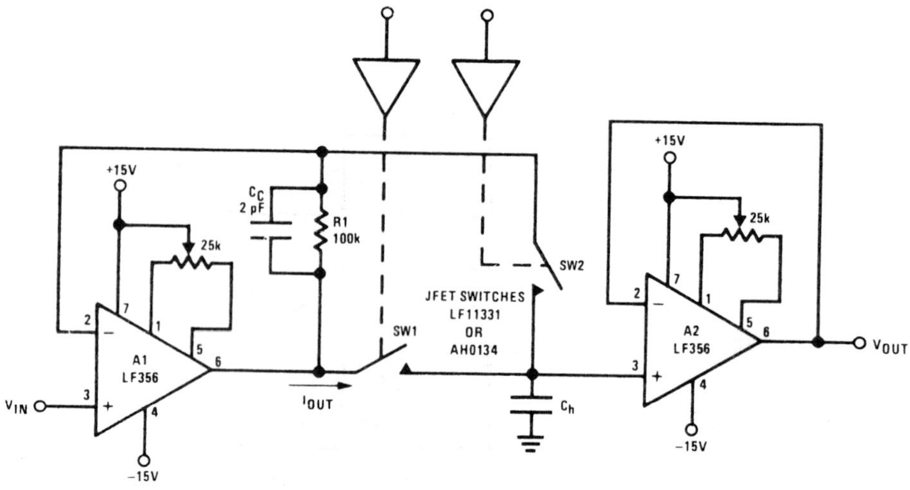

**High Accuracy Sample and Hold**

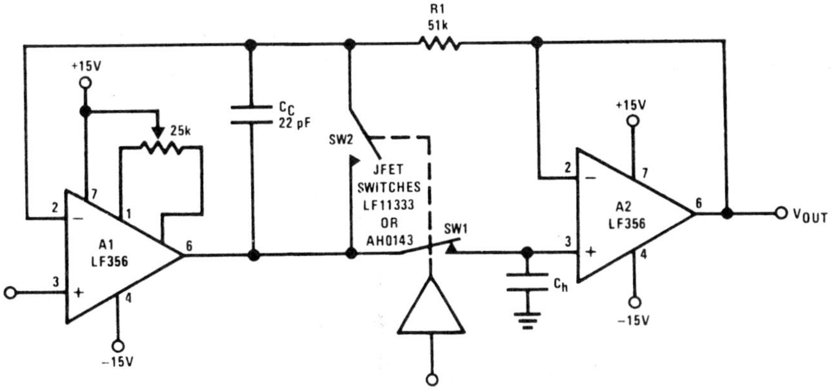


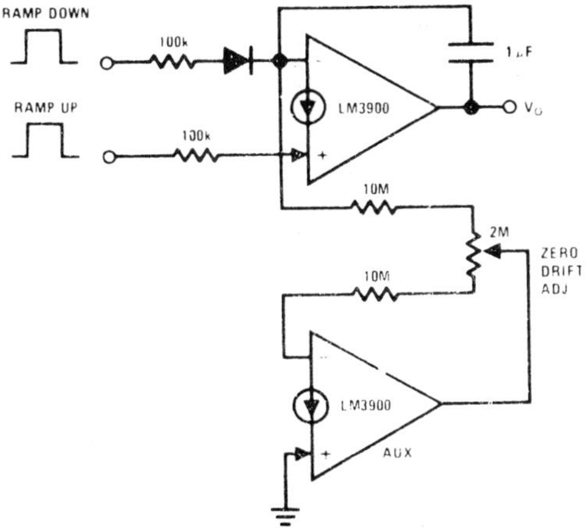


**Low-Drift Ramp and Hold Circuit**

## TACHOMETERS

The schematics following are a series of tachometers. Circuits presented are frequency averaging, frequency differencing, and a general tachometer.

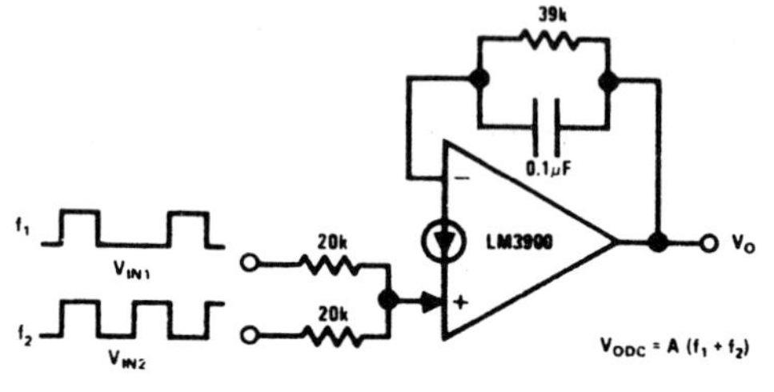


**Frequency Averaging Tachometer**

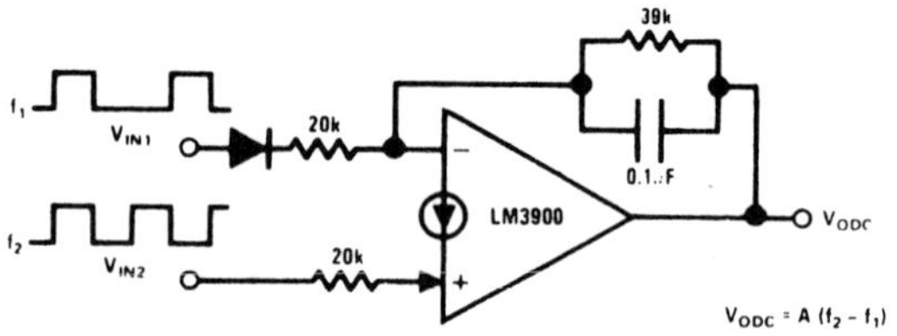


**Frequency Differencing Tachometer**

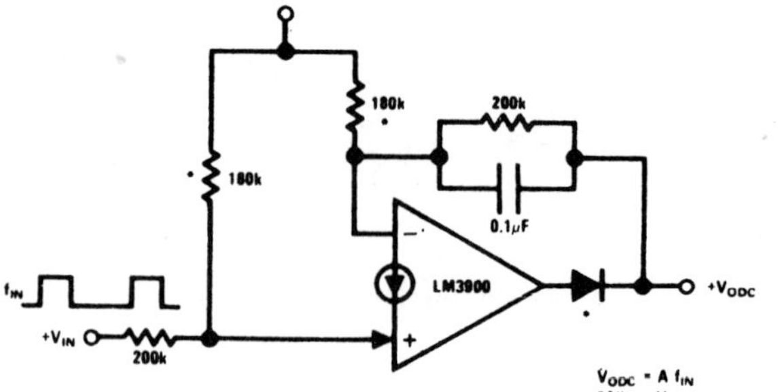


**Tachometer**

APPENDIX A

# Integrated-Circuit Manufacturers

**Advanced Micro Devices**
901 Thompson Place, P.O. Box 453, Sunnyvale, CA 94086
(408) 732-2400

**American Micro Systems, Inc. (AMI)**
3800 Homestead Road, Santa Clara, CA 95051
(408) 246-0330

**Analog Devices**
Route 1 Industrial Park, P.O. Box 280, Norwood, MA 02062
(617) 329-4700

**Beckman Instruments**
2500 Harbor Boulevard, Fullerton, CA 92634
(714) 993-8800

**Burr-Brown**
International Airport Industrial Park, Tucson, AZ 85734
(602) 746-1111

**Cherry Semiconductor**
99 Bald Hill Road, Cranston, RI 02920
(401) 463-6000

**Creatine Micro Systems**
11642-8 Knott Avenue, Garden Grove, CA 92641
(714) 898-9669

**Datel-Intersil**
11 Cabot Boulevard, Mansfield, MA 02048
(617) 339-9341

**Datricon**
7911 N.E. 33rd Drive, Suite 200, Portland, OR 97211
(503) 284-8277

**EMM/SESCO**
20630 Plummer Street, P.O. Box 668, Chatsworth, CA 91311
(213) 998-9090

**Exar Integrated Systems**
750 Palomar Avenue, Sunnyvale, CA 94086
(408) 732-7970

**Fairchild**
464 Ellis Street, Mountain View, CA 94042
(415) 962-5011

**Ferranti Electric, Inc.**
87 Modular Avenue, Commack, NY 11725
(516) 543-0200

**Fujitsu America, Inc.**
910 Sherwood, Lake Bluff, IL 60044
(312) 295-2610

**GTE Microcircuits**
2000 W. 14th Street, Tempe, AZ 85281
(602) 968-4431

**Harris Semiconductor**
P.O. Box 883, Melbourne, FL 32901
(305) 724-7000

**Hitachi**
707 W. Algonquin Road, Arlington Heights, IL 60005
(312) 593-7660

**Holt, Inc.**
3303 Harbor Boulevard, Suite D-5, Costa Mesa, CA 92626
(714) 754-1844

**Intel**
3065 Bowers Avenue, Santa Clara, CA 95051
(408) 987-8080

**Interdesign**
1255 Reamwood Avenue, Sunnyvale, CA 94086
(408) 734-8666

**Intersil**
10900 N. Tantau Avenue, Cupertino, CA 95014
(408) 996-5000

**LSI Computer Systems**
1235 Walt Whitman Road, Melville, NY 11747
(512) 271-0400

**Micro Power Systems**
3100 Alfred Street, Santa Clara, CA 95050
(408) 727-5350

**Mitel Semiconductor**
P.O. Box 13089, Kanata, Ottawa, Ontario, Canada K2K 1X3
(613) 592-2122

**Mitsubishi Electric Corp.**
Mitsubishi Denki Building, Marunouchi, Tokyo, Japan
(03) 218-3473

**Monolithic Memories, Inc.**
1165 East Arques Avenue, Sunnyvale, CA 94086
(408) 739-3535

**Mostek**
1215 West Crosby Road, Carrollton, TX 75006
(214) 323-6000

**Motorola Semiconductor Products**
5005 East McDowell Road, Phoenix, AZ 85008
(602) 244-6900

**National Semiconductor**
2900 Semiconductor Drive, Santa Clara, CA 95051
(408) 737-5000

**Nitron**
10420 Bubb Road, Cupertino, CA 95014
(408) 255-7550

**Omnibyte Corp.**
24W W. Roosevelt Road, Building I-S, West Chicago, IL 60185
(312) 231-6880

**Plessey Semiconductors**
1641 Kaiser Avenue, Santa Ana, CA 92715
(714) 540-9979

**Precision Monolithics, Inc.**
1500 Space Park Drive, Santa Clara, CA 95050
(408) 246-9222

**Raytheon Semiconductor**
350 Ellis Street, Mountain View, CA 94042
(415) 968-9211

**RCA Solid State Div.**
Box 3200, Somerville, NJ 08876
(201) 685-6000

**Rockwell International**
P.O. Box 3669, 3310 Miraloma Avenue, Anaheim, CA 92803
(714) 632-3729

**Semiprocess**
1885 Norman Street, Santa Clara, CA 95050
(408) 988-4004

**Signetics**
811 East Arques Avenue, P.O. Box 409, Sunnyvale, CA 94086
(408) 739-7700

**Silicon General**
11651 Monarch Street, Garden Grove, CA 92641
(714) 892-5531

**Siliconix, Inc.**
2201 Laurelwood Avenue, Santa Clara, CA 95054
(408) 988-8000

**Synertek**
3001 Stender Way, P.O. Box 552, Santa Clara, CA 95052
(408) 988-5600

**Tektronix**
P.O. Box 500, Beaverton, OR 97005
(503) 644-0161

**Teledyne Semiconductor**
1300 Terra Bella Avenue, Mountain View, CA 94043
(415) 968-9241

**Texas Instruments**
P.O. Box 225012, Dallas, TX 75265
(214) 238-6611

**Toshiba**
2151 Michelson Drive, Suite 190, Irvine, CA 92715
(714) 955-1155

**Universal Semiconductor**
1925 Zanker Road, San Jose, CA 95112
(408) 279-2862

**Zilog**
10460 Bubb Road, Cupertino, CA 95014
(408) 446-4666

APPENDIX B

# Operational Amplifier Manufacturers

| *Manufacturer* | *Prefix to Part No.* |
|---|---|
| Advanced Micro Devices | None |
| American Microsystems, Inc. | S |
| Analog Devices | AD |
| Burr-Brown | None |
| Datel | AM |
| Exar | XR |
| Fairchild | $\mu$A |
| Ferranti | ZN |
| Harris | HA |
| Intersil | ICL |
| Micro Power | MP |
| Motorola | MC |
| National | LF, LH, LM |
| Plessey | SL |
| Precision Monolithics, Inc. | OP or PM |
| Raytheon | RC |
| RCA | CA |
| Siemens | TAA |
| Signetics | SE or NE |
| Silicon General | SG |
| Siliconix | L or DF |
| Teledyne | None |
| Texas Instruments | TL |
| Toshiba | TA |

APPENDIX C

# Operational Amplifiers

## TYPICAL PART NUMBERS BY PURPOSE/FUNCTION

**General Purpose**

EXTERNAL COMPENSATION

| | |
|---|---|
| AD101A | Analog Devices |
| LM101A | National |
| CA101A | RCA |
| TA7506 | Toshiba |

INTERNAL COMPENSATION

| | |
|---|---|
| AD741 | Analog Devices |
| μA741 | Fairchild |
| OP-02 | PMI |
| TDA0741 | Signetics |

DUAL AMPLIFIERS

| | |
|---|---|
| XR1458 | Exar |
| RC1458 | Raytheon |
| SG1458 | Silicon General |

FET INPUT

| | |
|---|---|
| μAF155A | Fairchild |
| LF155A | Intersil |
| PM155A | PMI |

**High Output Current**

| | |
|---|---|
| AM-853OR | Datel |
| 3571A | Burr-Brown |
| LH0061 | National |
| CA3094 | RCA |
| TAA761 | Siemens |
| HA-2620 | Harris |

**High Speed**

| | |
|---|---|
| AM-500GC | Datel |
| LH0032 | National |
| HA-2535 | Harris |
| AD509K | Analog Devices |

**High Voltage**

| | |
|---|---|
| 3583 | Burr-Brown |
| MC1536 | Motorola |
| LM143 | National |
| SG1536 | Silicon General |
| ICH85101 | Intersil |

**Low Bias Current**

| | |
|---|---|
| ICH8500A | Intersil |
| AD515L | Analog Devices |
| 3523K | Burr-Brown |
| LH0052C | National |
| 1425-01 | Teledyne |

**Low Drift**

| | |
|---|---|
| AM-490-2C | Datel |
| AD504M | Analog Devices |
| MP5505P | Micro Power |
| LH0044A | National |
| μA714 | Fairchild |
| OP-05E | PMI |
| 3510A | Burr-Brown |

**Low Power**

| | |
|---|---|
| $\mu$A776 | Fairchild |
| HA-2720 | Harris |
| ICL4250 | Intersil |
| SG108 | Silicon General |
| AM-7611C | Datel |

**Programmable**

| | |
|---|---|
| SINGLE | |
| $\mu$A776C | Fairchild |
| ICL4250 | Intersil |
| MC1776 | Motorola |
| CA3078 | RCA |
| DUAL | |
| HA-2730 | Harris |
| ICL8022C | Intersil |
| LH24250 | National |
| SG1252 | Silicon General |
| TRIPLE | |
| ICL7631C | Intersil |
| CA3060 | RCA |
| SG1253 | Silicon General |
| L144A | Siliconix |
| QUAD | |
| XR146 | Exar |
| MC3575 | Motorola |
| LM146 | National |
| LM346 | Raytheon |

**Single Supply**

| | |
|---|---|
| SINGLE | |
| ICL7612C | Intersil |
| OP-20B | PMI |
| CA3130 | RCA |
| $\mu$A759 | Fairchild |
| DUAL | |
| OP-215B | PMI |
| CA158 | RCA |
| NE535 | Signetics |
| $\mu$A798 | Fairchild |
| QUAD | |
| LM3401 | National |
| CA3401 | RCA |
| $\mu$A124 | Fairchild |
| SG124 | Silicon General |
| XR3403 | Exar |
| OP-420B | PMI |

**Wide Band**

| | |
|---|---|
| 1435 | Teledyne |
| AM500GC | Datel |
| ADLH0032 | Analog Devices |
| $\mu$A715C | Fairchild |
| CA3015 | RCA |
| SL702C | Plessey |
| TL702M | Texas Instruments |

**Instrumentation**

| | |
|---|---|
| AD521J | Analog Devices |
| 3626A | Burr-Brown |
| AM-542AC | Datel |
| LH0036 | National |

APPENDIX D

# Typical Operational Amplifier Characteristics

| *Device* *Number* | *Offset Voltage* mV(25°) | *Bias Current* nA(25°) | *Offset Current* nA(25°) | *Voltage Drift* μV°C | *Band-width* MHZ | *Slew Rate* V/μS | *Voltage Gain* V/V | *CMRR* dB |
|---|---|---|---|---|---|---|---|---|
| **General Purpose** | | | | | | | | |
| *External Compensation* | | | | | | | | |
| AD101A | 2.0 | 75 | 10 | 15 | 1 | 0.5 | 50 K | 80 |
| *Internal Compensation* | | | | | | | | |
| μA741 | 5.0 | 500 | 200 | 3 | 1 | 0.5 | 50 K | 70 |
| *Dual Amplifier* | | | | | | | | |
| XR1458 | 6.0 | 500 | 200 | — | 1 | 0.5 | 20 K–50 K | 70 |
| *FET Input* | | | | | | | | |
| μAF155A | 2.0 | 0.050 | 0.01 | 5 | 2.5 | 3 | 50 K | 85 |
| **High Output Current** | | | | | | | | |
| 3571A | 2.0 | 0.10 | 0.05 | 40 | 0.5 | 3 | 50 K | 80 |
| **High Speed** | | | | | | | | |
| AM-500GC | 0.5 | 1.0 | 0.5 | 5 | 100 | 1,000 | 1,000 K | 80 |
| **High Voltage** | | | | | | | | |
| 3583 | 3 | 0.020 | 0.020 | 25 | 3 | 20 | 200 K | 110 |
| **Low Bias Current** | | | | | | | | |
| ICH8500A | 50 | 0.00001 | — | — | — | 0.5 | 20 K | 60 |

| *Device* *Number* | *Offset Voltage* mV(25°) | *Bias Current* nA(25°) | *Offset Current* nA(25°) | *Voltage Drift* μV°C | *Band-width* MHz | *Slew Rate* V/μS | *Voltage Gain* V/V | *CMRR* dB |
|---|---|---|---|---|---|---|---|---|
| **Low Drift** | | | | | | | | |
| AD504M | 0.5 | 80 | 10 | 0.5 | 0.3 | 0.12 | 1,000 K | 110 |
| **Low Power** | | | | | | | | |
| μA776C | 6.0 | 10 | 6 | 3 | 1 | 15 | 50 K | 70 |
| **Programmable** | | | | | | | | |
| *Single* | | | | | | | | |
| CA3078 | 45 | 170 | 32 | 6 | — | 0.4 | 25 K | 80 |
| *Dual* | | | | | | | | |
| HA-2730 | 3.0 | 5 | 3 | 10 | 25 | 0.1 | 40 K | 80 |
| *Triple* | | | | | | | | |
| ICL7631BC | 5.0 | 0.050 | 0.030 | 10 | 1 | 1.6 | 80 K | 70 |
| *Quad* | | | | | | | | |
| XR146 | 5.0 | 100 | 10 | — | 1 | 1.4 | 50 K | 70 |
| **Single Supply** | | | | | | | | |
| *Single* | | | | | | | | |
| ILC7612C | 15 | 0.050 | 0.030 | 25 | 1 | 1.6 | 80 K | 70 |
| *Dual* | | | | | | | | |
| OP-215B | 2.0 | 100 | 20 | 10 | 5.7 | 16 | 220 K | 100 |
| *Quad* | | | | | | | | |
| LM3401 | 20.0 | 300 | — | — | 5 | 0.6 | 1 K | — |
| **Wide Band** | | | | | | | | |
| 1435 | 5.0 | 20,000 | 300 | 25 | 500 | 250 | 10 K | 80 |
| **Instrumentation** | | | | | | | | |
| AD522A | 0.40 | 25 | 20 | 6 | 0.3 | 0.1 | 1–1,000 | 75 |

APPENDIX E

# IC Manufacturers' Logos

| LOGO | MANUFACTURER |
|---|---|
| AM* | ADVANCED MICRO DEVICES |
| AMI Ai* | AMERICAN MICRO SYSTEMS |
| ANALOG DEVICES | ANALOG DEVICES |
| ANALOGIC | ANALOGIC |
| BECKMAN Beckman* | BECKMAN INSTRUMENTS |
| BURR-BROWN BB | BURR-BROWN |
| CS MCC | CHERRY SEMICONDUCTOR (Micro Components) |
| MOS | COMMODORE MOS TECHNOLOGY |
| | CONSUMER MICROCIRCUITS |
| CY* | CYBERNETIC MICRO DEVICES* |
| DDC | DATA DEVICE CORPORATION |
| | DATA GENERAL |
| | DATEL-INTERSIL |
| | ELECTRONIC ARRAYS |
| | EXAR INTEGRATED SYSTEMS |
| F FAIRCHILD F* | FAIRCHILD |
| FERRANTI | FERRANTI |
| F * FUJITSU LIMITED | FUJITSU |
| GI | GENERAL INSTRUMENT |

| LOGO | MANUFACTURER |
|---|---|
| HUGHES | HUGHES AIRCRAFT |
| Hybrid Systems * Hybrid Systems | HYBRID SYSTEMS |
| | INTEGRATED PHOTOMATRIX |
| intel i | INTEL |
| interdesign | INTERDESIGN |
| INTERSIL | INTERSIL |
| | ITT semiconductors |
| LSI | LSI COMPUTER SYSTEMS |
| | LAMBDA |
| ML | MASTER LOGIC |
| | MICRO NETWORKS |
| Mii | MICROPAC INDUSTRIES |
| | MICRO POWER SYSTEMS |
| | MITEL SEMICONDUCTOR |
| | MITSUBISHI |
| MMI MMI* | MONOLITHIC MEMORIES |
| MOSTEK | MOSTEK |
| | MOTOROLA |
| NS* | NATIONAL SEMICONDUCTOR |

| LOGO | MANUFACTURER |
|---|---|
| GTE | GTE MICROCIRCUITS |
| | HARRIS |
| | HITACHI |
| Nitron<br>NI INC* | NITRON |
| NORTEC | NORTEC |
| OKI® | OKI |
| oei<br>OEI* | OPTICAL ELECTRONICS |
| MATSUSHITA | PANASONIC (Matsushita) |
| PLESSEY | PLESSEY |
| PMI<br>PMI* | PRECISION MONOLITHICS |
| RCA Solid State | RCA |
| RAY | RAYTHEON SEMICONDUCTOR |
| RETICON® | RETICON |
| Rockwell | ROCKWELL (COLLINS) |
| <br>SGS ATES* | SGS-ATES SEMICONDUCTOR |
| | SANYO |
| signetics<br>S* | SIGNETICS |
| SG<br>Silicon General | SILICON GENERAL |
| SSi | SILICON SYSTEMS |
| | SILICONIX |
| SSS | SOLID STATE SCIENTIFIC |
| NCR | NCR |
| NEC | NEC |
| S S* | SOLITRON |
| SPRAGUE | SPRAGUE ELECTRIC |
| | STANDARD MICROSYSTEMS |
| | SUPERTEX |
| | SYNERTEK |
| TELEDYNE CRYSTALONICS | TELEDYNE CRYSTALONICS |
| TELEDYNE PHILBRICK | TELEDYNE PHILBRICK |
| | TELEDYNE SEMICONDUCTOR |
| TELE FUN KEN TFK | TELEFUNKEN |
| TI | TEXAS INSTRUMENTS |
| | THOMSON-CSF COMPONENTS |
| TMX | TMX |
| Toshiba | TOSHIBA |
| TRW | TRW |
| | UNIVERSAL SEMICONDUCTOR |
| | WESTERN DIGITAL |
| Zilog | ZILOG |

APPENDIX F

# Bibliography


Coughlin, Robert F., and Frederick F. Driscoll. *Operational Amplifiers and Linear Integrated Circuits.* Englewood Cliffs, N.J.: Prentice-Hall, Inc., 1977.

Faulkenberry, Luces M. *An Introduction to Operational Amplifiers.* New York: John Wiley & Sons, Inc., 1977.

Gise, Peter E., and Richard Blanchard. *Semiconductor and Integrated Circuit Fabrication Techniques.* Reston, Va.: Reston Publishing Co., 1979.

Graeme, Jerald G. *Applications of Operational Amplifiers.* New York: McGraw-Hill Book Co., 1973.

Graeme, Jerald G., Gene E. Tobey, and Lawrence P. Huelsman. *Operational Amplifiers.* New York: McGraw-Hill Book Co., 1971.

Hoenig, Stuart A. *How to Build and Use Electronic Devices Without Frustration, Panic, Mountains of Money, or an Engineering Degree.* Boston: Little, Brown & Co., 1980.

Jung, Walter G. *IC Op-Amp Cookbook.* Indianapolis, In.: Howard W. Sams & Co., Inc., 1980.

Roberge, James K. *Operational Amplifiers, Theory and Practice.* New York: John Wiley & Sons, Inc., 1975.

Seippel, Robert G. *Designing Circuits with IC Operational Amplifiers.* Chicago: American Technical Society, 1975.

Sheingold, Daniel H. *Transducer Interfacing Handbook.* Norwood, Mass.: Analog Devices, Inc., 1981.

Wojslaw, Charles. *Integrated Circuits.* Reston, Va.: Reston Publishing Co., 1978.

———. *Electronic Concepts, Principles, and Circuits.* Reston, Va.: Reston Publishing Co., 1980.

# Index

10 363 348 01